A Focus on Multiplica̶

A Focus on Multiplication and Division is a groundbreaking effort to make mathematics education research readily accessible and understandable to pre- and in-service K–6 mathematics educators. Revealing students' thought processes with extensive annotated samples of student work and vignettes characteristic of teachers' experiences, this book is sure to equip educators with the knowledge and tools needed to modify their lessons and to improve student learning of multiplication and division.

Special Features:

- *Looking Back Questions* at the end of each chapter allow teachers to analyze student thinking and to consider instructional strategies for their own students.
- *Instructional Links* help teachers relate concepts from each chapter to their own instructional materials and programs.
- *Big Ideas* frame the chapters and provide a platform for meaningful exploration of the teaching of multiplication and division.
- *Answer Key* posted online offers extensive explanations of in-chapter questions.

Each chapter includes sections on the Common Core State Standards for Mathematics and integrates the Ongoing Assessment Project (OGAP) Multiplicative Reasoning Progression for formative assessment purposes. Centered on the question of how students develop their understanding of mathematical concepts, this innovative book places math teachers in the mode of ongoing action researchers.

Studies in Mathematical Thinking and Learning

Alan H. Schoenfeld, Series Editor

A Focus on Multiplication and Division

Bringing Research to the Classroom

Elizabeth T. Hulbert, Marjorie M. Petit, Caroline B. Ebby, Elizabeth P. Cunningham, and Robert E. Laird

Routledge
Taylor & Francis Group

NEW YORK AND LONDON

First published 2017
by Routledge
711 Third Avenue, New York, NY 10017

and by Routledge
2 Park Square, Milton Park, Abingdon, Oxon, OX14 4RN

Routledge is an imprint of the Taylor & Francis Group, an informa business

© 2017 Taylor & Francis

The right of Elizabeth T. Hulbert, Marjorie M. Petit, Caroline B. Ebby, Elizabeth P. Cunningham, and Robert E. Laird to be identified as authors of this work has been asserted by them in accordance with sections 77 and 78 of the Copyright, Designs and Patents Act 1988.

Library of Congress Cataloging-in-Publication Data
Names: Hulbert, Elizabeth T.
Title: A focus on multiplication and division : bringing research to the
  classroom / by Elizabeth T. Hulbert [and four others].
Description: New York : Routledge, 2017. | Includes
  bibliographical references and index.
Identifiers: LCCN 2017003234 | ISBN 9781138205680
  (hardback) | ISBN 9781138205697 (pbk.)
Subjects: LCSH: Mathematics teachers—Training of. | Elementary
  school teachers—Training of. | Multiplication—Study and teaching
  (Elementary) | Division—Study and teaching (Elementary)
Classification: LCC QA10.5 .F63 2017 | DDC 372.7/2—dc23
LC record available at https://lccn.loc.gov/2017003234

ISBN: 978-1-138-20568-0 (hbk)
ISBN: 978-1-138-20569-7 (pbk)
ISBN: 978-1-315-16361-1 (ebk)

Typeset in Minion
by Apex CoVantage, LLC

Visit the eResources: www.routledge.com/9781138205697

This book is dedicated to the thousands of teachers who participated in Ongoing Assessment Project (OGAP) work during the past 14 years. The student solutions and suggestions they shared affected this book in countless ways. We thank them for their engagement and enthusiasm.

Contents

Contents

Preface

The Importance of Multiplicative Reasoning

Multiplicative reasoning is a central focus of elementary-grade mathematics. In fact, the ability to reason with a variety of multiplicative concepts and flexibly use multiplicative skills, strategies, and algorithms to solve problems are prerequisites for future work with foundational mathematical ideas such as fractions, decimals, percent, ratios, and proportions (Siemon, Breed, & Virgona 2005). Some researchers believe that the transition from additive to multiplicative thinking is a major barrier to learning the important mathematics in the middle grades (Siemon et al., 2005). Because of the importance of multiplicative reasoning, mathematics instruction in the elementary years must focus attention on the overarching ideas of place-value, multiplicative thinking, and rational number ideas in order to prepare students to progress to the next "big idea" (Siemon & Virgona 2001).

A Focus on Multiplication and Division: Bringing Research to the Classroom

A Focus on Multiplication and Division: Bringing Research to the Classroom had its beginnings as a series of professional development sessions. These sessions were designed to help teachers understand the depth and breadth of multiplicative reasoning, the ways students learn these sometimes complex ideas, and strategies to both monitor student learning and make informed instructional decisions based on evidence in student solutions. These sessions were part of the Ongoing Assessment Project (OGAP), which began in 2003 with the goal of bringing formative assessment tools and strategies to mathematics teachers. OGAP is built on the mathematics education research on how students learn specific mathematical concepts related to proportionality, fractions, multiplicative reasoning, and additive reasoning.

It soon became obvious to us that the ideas and information contained in these sessions were important for all teachers charged with teaching multiplication and division, and *A Focus on Multiplication and Division* began to take form.

This book follows in the footsteps of its predecessor, *A Focus on Fractions: Bringing Research to the Classroom 2nd edition*, which has been well received by math teachers, special educators, administrators, math teacher educators, pre-service teachers, and mathematics professional development providers. The success of *A Focus on Fractions* solidified our beliefs that teachers want to become familiar with the robust body of mathematics education research about how students learn particular mathematical ideas, as well as concrete ways this knowledge can inform mathematics instruction. We have found that when teachers understand foundational mathematics education research, are provided tools and strategies for analyzing evidence of student thinking, and deeply understand the mathematics content central to their grade level, they thoughtfully and intentionally change their instruction in ways that support deeper student learning. At its core, *A Focus on Multiplication and Division: Bringing Research to the Classroom* is an attempt to bring this knowledge to teachers.

Central Features of This Book

The *OGAP Multiplicative Framework* plays a prominent role in *A Focus on Multiplication and Division*. This framework interprets and clearly communicates the mathematics education research on how students learn multiplicative reasoning concepts in ways that teachers can use to improve their instruction and their students' understanding. The framework, introduced in detail in Chapter 2 and returned to again and again throughout the book, is a tool to help teachers better understand and use activities and lessons in the textbook, select or design formative assessment tasks, understand evidence in student work, make instructional decisions, and provide actionable feedback to students. We continue to find that most mathematics teachers are unfamiliar with this research despite the fact that much of it has been available for years. The *OGAP Multiplicative Framework* is our attempt to share the knowledge of mathematics education researchers with mathematics teachers in a clear, concise, and understandable way so that this type of research comes out of the shadows and into teachers' practices. The *OGAP Multiplication and Division Progressions* are key components of the framework and communicate how students develop understanding of multiplication and division concepts, strategies, and procedures, as well as the common errors students make and misconceptions that affect new learning. We know that most teachers overwhelmingly value this research and immediately appreciate how it can improve the effectiveness of their instruction.

This book uses over 100 examples of authentic student work to communicate specific research ideas, as well as to help readers understand particular mathematical concepts. The work samples contained in *A Focus on Multiplication and Division* have been collected from OGAP pilot studies and through our ongoing work with teachers across the country receiving OGAP professional development. Over the years we have collected thousands of pieces of student work, and we cannot overstate how much we have learned about the teaching and learning of multiplicative concepts through our analysis and discussion of these solutions. We include a wide variety of student work samples in this book with the hope that through your interaction with them you, too, gain a deep and robust understanding of important aspects of teaching and learning multiplication and division skills and concepts.

Most chapters contain a brief discussion of the ways in which the Common Core State Standards in Mathematics (CCSSM) are supported by the research base that underpins *A Focus on Multiplication and Division*.

Near the end of each chapter you will find a section titled *Looking Back* that poses questions designed for the reader to more deeply examine particular ideas posed in the chapter or to contemplate related concepts. The answers to these questions are found at www.routledge.com. Most chapters conclude with an *Instructional Link: Your Turn*. This section generally asks the reader to analyze her or his own instruction and mathematics program in light of the key ideas presented in the chapter. It is our hope that readers use the *Instructional Link* to help them make thoughtful and intentional instructional adaptations consistent with the important concepts in the book.

A Book for Teachers

A Focus on Multiplication and Division: Bringing Research to the Classroom is primarily written for classroom and preservice teachers. The mathematics content related to multiplication and division, the authentic student work samples, and the *Instructional Link* and *Looking Back* sections are specifically designed to help teachers learn and reflect on the foundational mathematics education research and ways to use this knowledge to analyze student thinking, take action based on the evidence in their students' solutions, and use textbook materials more effectively. Teams of math teachers at the same grade level, from the same school, or involved in professional development related to multiplication and division can benefit greatly from reading and discussing the chapters and answering and discussing questions posed in *Looking Back* and the *Instructional Links*. This type of engagement with *A Focus on Multiplication and Division* works well in professional learning communities (PLC) in place in many schools.

The numerous samples of authentic student work can be invaluable to instructors working with preservice teachers, as preservice teachers often do not have access to authentic student work. In addition, *A Focus on Multiplication and Division* provides preservice teachers with an introduction to important educational research related to multiplication and division, research that is vital, yet lacking, for many teachers.

Final Thoughts

There are many important and thought-provoking ideas in this book, yet the "heart and soul" of *A Focus on Multiplication and Division* is the mathematics education research related to the teaching and learning of whole number multiplication and division and the important contributions that analyzing student work make to effective and informed instruction. The ability to analyze student work to understand how students understand mathematical concepts and skills is paramount if we are to pose the right questions and design effective lessons that provide the best opportunities for all students to learn important mathematics. We hope *A Focus on Multiplication and Division* helps strengthen these abilities and makes them an integral part of teachers' instruction.

Acknowledgments

We extend our deep appreciation to the thousands of Vermont, New Hampshire, Alabama, Nebraska, Michigan, Charleston, and Philadelphia teachers who have been involved in the Ongoing Assessment Project (OGAP) over the past 14 years. We also thank the original OGAP design team members who were instrumental in the early years of the project, as well as the members of the current OGAP National Professional Development Team.

Additionally, we extend our sincere thanks to Fritz Mosher, Caroline Ebby, and Jon Supovitz from the Consortium for Policy and Research in Education at Teachers College Columbia University and the University of Pennsylvania for their ongoing support and guidance as OGAP evolved over the past decade.

Also: Our thanks to Bridget Goldhahn for her assistance with much of the line art and Christopher Cunningham for his Go To artwork.

The student work samples in this book were collected over the past decade through the Ongoing Assessment Project, which began in 2003 as part of the Vermont Mathematics Partnership.

1

What Is Multiplicative Reasoning?

> ### Big Ideas
>
> - Although students may use repeated addition or subtraction to solve multiplication or division problems, multiplication and division are not simply an extension of addition and subtraction.
> - Multiplicative reasoning is about understanding situations where multiplication or division is an appropriate operation and having a variety of skills and concepts to approach those situations flexibly.
> - There are a number of skills and concepts that a student must understand in order to be fluent with multiplication and division.
> - Strong multiplicative reasoning involving whole numbers provides the foundation for fractional and proportional reasoning.

Multiplicative reasoning is a cornerstone to success in other mathematical topics and a potential gatekeeper to success both in and out of school. Multiplicative reasoning is foundational for the understanding of many of the mathematical concepts that are encountered later in students' school career, such as ratios, fractions, and linear functions, as well as in everyday situations (Vergnaud, 1983). In this chapter multiplicative reasoning is described in two ways: 1) mathematically and 2) from a teaching and learning perspective.

Multiplicative Reasoning: The Mathematics

"Multiplicative reasoning is a complicated topic because it takes different forms and deals with many different situations" (Nunes & Bryant, 1996, p. 143). In elementary school, for example, students engage in a range of multiplication and division contexts, including equal groups, equal measures, unit rates, measurement conversions, multiplicative comparisons, scaling, and area and volume. The types of quantities involved and how the quantities interact are key to understanding multiplication and division in these different contexts.

1

In this chapter the general concept of initial multiplicative understanding is examined. Chapter 5 provides an in-depth discussion of each of these contexts and how they affect student strategies and reasoning.

This section focuses on the following:

1. Why multiplication and division are not simply an extension of addition and subtraction:
   - How the number relationships in addition are different from those in multiplication.
   - How the actions in addition are different from those in multiplication.
2. The difference between additive (absolute) reasoning and multiplicative (relative) reasoning.

There is a commonly held belief among many educators that multiplication and division are just extensions of addition and subtraction. This belief arises because it is possible to solve whole number multiplication and division problems using repeated addition and subtraction, respectively. However, multiplication and division involve a different set of number relationships and different actions than addition and subtraction, which are described next.

Number Relationships and Actions in Addition

Additive reasoning involves situations in which sets of objects are joined, separated, or compared. For example, 3 apples + 6 apples = 9 apples. Each apple is a separate entity, and the sum is the union of all the apples as shown in Figure 1.1. It is also important to remember that in additive situations the numbers tell the actual size of each set. So in the case of 3 apples + 6 apples, the 3 means how many in one set and the 6 means how many in another set. The numbers in additive situations represent the value of each of the independent sets and do not rely on the other number for meaning or value. As you will see, this is not the case with multiplication and division.

Figure 1.1 The sum of 3 apples and 6 apples is 9 apples (3 apples + 6 apples = 9 apples).

Number Relationships and Actions in Multiplication

One major difference between additive reasoning and multiplicative reasoning is that multiplicative reasoning does not involve the actions of joining and separating, but instead, it often involves the action of iterating or making multiple copies, of a unit. Instead of involving one-to-one correspondence (1 apple) as in additive reasoning, multiplication often involves many-to-one correspondence (Nunes & Bryant, 1996). It is important to note that the examples used in this section are equal group problems, one of the first many-to-one situations that elementary students encounter in instruction.

To understand what is meant by this many-to-one correspondence, consider a plate with 3 apples. One plate has 3 apples, so the many-to-one relationship is 3 apples to 1 plate. When you increase the number of plates, you increase the number of apples by the number of apples on each plate. The constant relationship of 3 apples to 1 plate, however, never changes. When thinking about answering the question how many apples on 6 plates, you can think of this as 6 iterations of 3 apples to 1 plate. Figure 1.2 shows the 3 apples per plate iterated 6 times (once for each plate of 3 apples).

6 plates × 3 apples on each plate = 18 apples

[handwritten: 6 + 6 3 × 1 3 × 2 1 to 1 many to 1]

Figure 1.2 Six times 3 apples on each plate equals 18 apples. That is, the composite unit of 3 apples to one plate is iterated 6 times (6 plates × 3 apples in each plate = 18 apples).

3 apples to a plate
Composite unit

[handwritten margin notes: Vocab; Scalar factor; Composite Unit; Unitizing; How many/much Something is increasing by]

Notice that the number of apples is scaled up or down depending on the number of plates. The number of iterations of the composite unit (e.g., 3 apples to each plate; 3 apples per plate) called for in a problem can also be thought of as the scalar factor. In the case of Figure 1.2, the scalar factor is 6, meaning that the composite unit of 3 (apples per plate) is iterated 6 times, resulting in a product of 18 apples.

Researchers refer to the many-to-one relationship as the composite unit. Students who see, iterate, and operate with the composite unit are unitizing. Unitizing refers to the understanding that quantities can be grouped and then the group can be referred to as one unit yet have a value greater than 1. Imagine a package of 4 cookies. The unit is the package, but the package has a value of 4 cookies. There is a more detailed discussion of this idea in Chapter 4. Conceptually it is harder for children to mentally keep track of a composite unit than it is to count by ones, as they must coordinate two levels of units (the composite unit and the number of units you count by) (Steffe, 1992; Ulrich, 2015).

Students who first begin to unitize may solve multiplication problems by iterating the composite unit using repeated addition instead of multiplication. This evidence often leads one to think that multiplication is an extension of addition. However, what distinguishes repeated addition in a multiplicative situation from additive reasoning is the composite unit (many-to-one) is iterated and added. When students develop more sophisticated multiplicative reasoning, they can conceptualize 18 as being made up of 6 composite units of 3 individual units each.

As students develop their understanding and flexibly use unitizing, they will move away from the repeated addition strategy to strategies that involve multiplication. Consider Lyla's explanation:

> "Well, I know that 8×8 is 64 so to find 8×7 I just need to take 1 away—I mean 1 row of 8 away—and that would be 56."

In explaining her strategy for finding the product of 8 and 7, Lyla uses language that indicates an understanding of unitizing. Note Lyla's reference to removing a row. Iterations of a composite unit can take many visual forms. Some of these may be one column or row in an array iterated multiple times or one group in a set iterated multiple times. Lyla's explanation helps to make sense of what happens when one of the numbers in a multiplication situation is changed. Changing the value of either the composite unit or the scalar factor has an impact on the total by the value of the other number.

Consider the situation of 6 plates with 3 apples per plate again. By decreasing the number of plates to 5, the number of apples decreases by one composite unit, or 1 plate of 3 apples. See Figure 1.3.

In multiplication the quantities are different from each other and yet dependent on each other. In this example the total number of apples is dependent on the number of plates. This understanding represents a significant difference between addition and multiplication.

Figure 1.3 Five plates × 3 apples in each plate is 3 apples (one composite unit) less than 6 plates × 3 apples in each plate.

6 plates x 3 apples on each plate = 18 apples

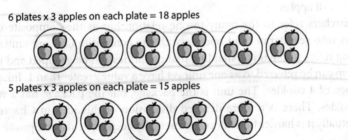

5 plates x 3 apples on each plate = 15 apples

CompositeUnit → Cant be broken down
Additive → Vale Can be moved around

Students who reason multiplicatively can unitize, or see a composite unit, and then create multiple copies of it. As stated earlier, however, this is just the beginning of considering the meaning of multiplication in a very complex landscape. Although initially an understanding of multiplication is reliant on students making multiple copies, or iterations, of that composite unit, eventually we need students to broaden their multiplicative reasoning to include more complex actions and contexts. These will be discussed in detail throughout this book. A summary of the difference in number relationships and actions between additive and multiplicative situations discussed in this section is found in Table 1.1.

Table 1.1 Differences between additive and multiplicative reasoning discussed in this section.

	Number Relationships	Actions
Additive Reasoning	One-one correspondence	Joining, separating, or comparing
Multiplicative Reasoning	Many-to-one composite unit	Iterating and scaling

Absolute and Relative Differences

Another way to think about the difference between additive reasoning and multiplicative reasoning is to consider the idea of absolute versus relative thinking (Lamon, 2012). Think about two pieces of rope lying on a table as shown in Figure 1.4. Rope A is 8 inches long and Rope B is 24 inches long. How do you think the students will describe how much longer one rope is than the other rope?

Figure 1.4 How much longer is Rope B than Rope A?

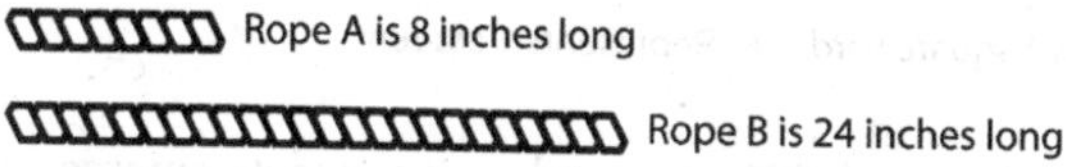

Rope A is 8 inches long

Rope B is 24 inches long

Read these responses from two students (Figure 1.4). How do they each think about the situation?

Tasha's response: Rope B is 16 inches longer than Rope A. → *additive/comparative, concrete*

Ben's response: Rope B is 3 times longer than Rope A. - *Multiplicative, abstract*

Tasha sees the length of Rope B as an absolute length independent of anything else so when comparing just finds the difference between the two lengths. This is an additive way of thinking about comparing these two lengths. Ben thinks about the comparative or relative perspective, which is a multiplicative way of thinking about answering the same question. Although both ways of thinking are correct and can be helpful in viewing situations, Ben's thinking is more abstract and is "necessary for moving beyond counting and absolute thinking" (Lamon, 2012, p. 31) in order to make sense of more complex situations. Before working with fractions and decimals, students need to expand their ability to include both absolute and relative thinking. They must see the world, and situations in the world, from a relative perspective.

The concept of "crowdedness" provides another way to think about the differences between absolute and relative thinking. Examine Figure 1.5. You'll notice that there are 10 more students in the same space during the sixth grade lunch period than during the seventh grade lunch period. This is an additive, or absolute, way to think about the situation. In contrast, thinking about this multiplicatively helps to describe the relative difference between the two situations. That is, there is 1 square yard per student during sixth grade lunch compared to 2 square yards per student during the seventh grade lunch. In this case it is easy to see that the seventh grade students have twice the room as the sixth grade students. Thus, the sixth grade lunch period is twice as crowded as the seventh grade lunch period.

Figure 1.5 From an absolute perspective there are 10 more students eating lunch during the sixth grade lunch period than during the seventh grade lunch period. From a relative perspective the seventh grade students have twice the space as the sixth grade students to eat lunch.

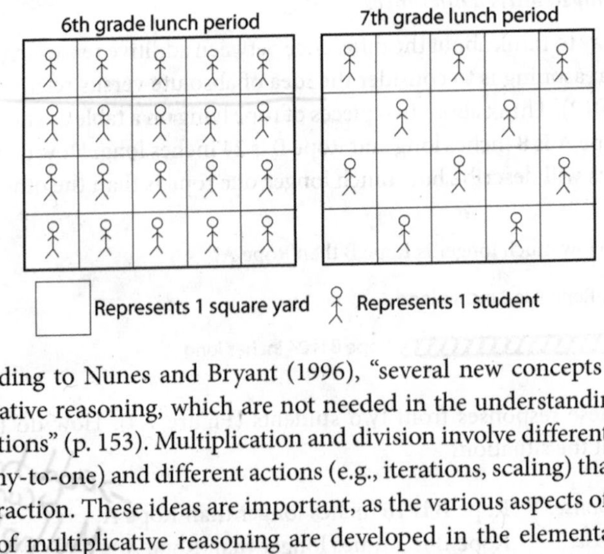

According to Nunes and Bryant (1996), "several new concepts emerge in multiplicative reasoning, which are not needed in the understanding of additive situations" (p. 153). Multiplication and division involve different quantities (e.g., many-to-one) and different actions (e.g., iterations, scaling) than addition and subtraction. These ideas are important, as the various aspects of the development of multiplicative reasoning are developed in the elementary grades. When looking closely at the various forms multiplication can take, it is important to appreciate that it is not simply a more complex version of addition.

Multiplicative Reasoning: From a Teaching and Learning Perspective

Although many teachers believe that multiplicative reasoning is confined to learning how to multiply and divide multidigit whole numbers and memorize math facts, this book will help you see this is just a small part of the picture. When thinking about the skills and concepts that help to support the development of strong multiplicative reasoning, it is important to consider both procedural fluency and conceptual understanding. Procedural fluency is more than applying the steps to an algorithm. Rather procedural fluency "refers to knowledge of procedures, knowledge of when and how to use them appropriately, and skill in performing them flexibly, accurately, and efficiently" (National Research Council [NRC], 2001, p. 121). Procedural fluency works together with conceptual understanding, each contributing to a deeper understanding of the other.

Conceptual understanding refers to an integrated and functional grasp of mathematical ideas. Students with conceptual understanding know more than isolated facts and methods. They understand why a mathematical idea is important and the kinds of contexts in which it is useful. (NRC, 2001, p. 110)

Examine the four pieces of student work in Figures 1.6 to 1.9. These are responses to tasks from the same fifth grade student. Use the work, as well as your experience as a teacher, to complete this sentence:

"The student who has multiplicative reasoning shows evidence of . . ."

Figure 1.6 Ethan solves equal groups problems using the traditional US algorithm.

(a) Mark bought 12 boxes of crayons. Each box contains 8 crayons. How many crayons were there altogether? Show your work.
(b) John bought 12 boxes of crayons. Each box contained 64 crayons. How many crayons were there altogether? Show your work.

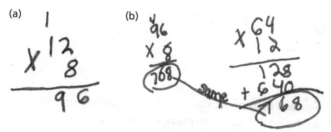

Figure 1.7 Ethan shows evidence of applying the partial products algorithm to solving a problem by recognizing that he still has to multiply 34 × 40 to complete the multiplication of 34 × 42.

Study this number sentence:

34 × 42

Sharon started to solve this problem by finding the answer to 34 × 2. What more does she need to do to get the answer to 34 × 42? Explain your thinking.

34x40

then when she finds the answer
34x40 and then add the two products
for 34x2 and 34x40

Figure 1.8 Ethan solves the problem using place-value understanding by decomposing 19 into 10 + 9 and the distributive property: $3 \times (10 + 9) = (3 \times 10) + (3 \times 9)$.

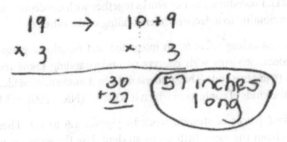

A piece of elastic stretches to 3 times its length. If the elastic is 19 inches long, how long will it be when it is fully stretched?

Figure 1.9 Ethan uses doubling and halving to solve this equal groups problem.

A class has set a goal that each student will read 45 pages this week. There are 16 students in the class. How many pages will they have read altogether by the end of the week?

$$45 \times 16 = ?$$
$$90 \times 8 = 720$$

 To learn more about the doubling and halving strategy go to Chapter 4: The Role of Concepts and Properties.

When looking across the four pieces of student work, there is evidence that Ethan understands the relationship between the numbers in the different problems and has a variety of efficient strategies to solve different kinds of word problems. In Figure 1.6 he uses the solution to part (a) to find the solution to part (b). In Figure 1.9 Ethan demonstrates understanding that there is a relationship between the factors and that the change to one factor affects treatment of the other factor by applying a doubling and halving strategy. This is another example of recognizing the relationship between the numbers in the tasks. Ethan uses two different algorithms within the four tasks, demonstrating flexibility with multidigit algorithms. He uses the traditional US algorithm in Figure 1.6, and in Figure 1.8 he uses partial products founded on the application of the distributive property to find the product of 19 and 3. In Figure 1.7 Ethan also demonstrates a deeper understanding of the distributive property by explaining that he recognizes the part of the task that is yet to be completed.

Based on the student work in Figures 1.6 to 1.9, you may have come up with all or some of these characteristics in response to the statement "the student who has multiplicative reasoning shows evidence of . . ."

- Recognizing multiplication as the operation to use in a variety of situations,
- Using a variety of strategies depending on the numbers and the number relationships,
- Understanding and using the properties of operations flexibly,
- Demonstrating fluency with multiplication facts,
- Solving multidigit multiplication problems efficiently.

Based on your own knowledge of students who demonstrate strong multiplicative reasoning, you may have thought of a number of other skills and concepts that were not observable in this student work. Some of these might be:

- Understands the inverse relationship between multiplication and division,
- Understands unitizing,
- Understands multiplication as equal grouping,
- Knows what the numbers mean in a multiplication situation,
- Uses and interacts with a variety of models to represent multiplication,
- Understands the meaning of remainders.

This is just the beginning of a long list of skills and concepts necessary for multiplicative reasoning, because developing strong multiplicative reasoning is a complex endeavor. These are the foundations of procedural fluency, which is built on conceptual understanding of multiplication and division. Over the course of this book, we will examine and consider all aspects of what it means to reason multiplicatively.

The *OGAP Multiplicative Framework* and Progressions

 The *Ongoing Assessment Project (OGAP) Multiplicative Framework* can be downloaded at www.routledge.com/9781138205697.

It is a synthesis of mathematics education research on how students develop multiplication and division fluency based on conceptual understanding as well as common errors that students make or preconceptions or misconceptions that interfere with learning new concepts or solving related problems.

Examine the *OGAP Multiplication and Division Progressions* on the centerfold of the *OGAP Multiplicative Framework*. The *OGAP Multiplication and Division Progressions* are representations of the development of the understanding and fluency necessary for multiplication and division. The progressions represent student strategies from least to most sophisticated, moving from bottom to top. Look at the progressions in relation to your list of characteristics of multiplicative reasoning, and consider where you see those characteristics represented. Are there additional characteristics that are not on your list? In Chapter 2 we will look at both the format and content of the *OGAP Multiplicative Framework* in detail. The purpose of this quick look is to see the complexity of the acquisition of multiplicative reasoning.

Go To   For more on the *OGAP Multiplicative Framework* go to Chapter 2: The OGAP Multiplication Progression.

It is important for teachers to know about the skills and concepts a student must possess in order to make instructional decisions, more effectively meet students' needs, and move students toward acquiring meaningful mathematical knowledge. This is a set of knowledge critical to teaching called *pedagogical content knowledge* (Shulman, 1986). According to Cochran (1991), "pedagogical content knowledge is a type of knowledge that is unique to teachers, and in fact what teaching is about" (p. 5).

As Shulman (1986) explains, pedagogical content knowledge:

> goes beyond knowledge of subject matter per se to the dimension of subject matter knowledge for teaching . . . Pedagogical content knowledge also includes an understanding of what makes the learning of specific topics easy or difficult: the conceptions and preconceptions that students of different ages and backgrounds bring with them to the learning of those most frequently taught topics and lessons. If those preconceptions are misconceptions, which they so often are, teachers need knowledge of the strategies most likely to be fruitful in reorganizing the understanding of learners, because those learners are unlikely to appear before them as blank slates (p. 9).

The *OGAP Multiplicative Framework* is one tool that can be used to enhance teachers' pedagogical content knowledge and help teachers make instructional decisions, as it illustrates the progression of skills and concepts students should experience and acquire as they become multiplicative reasoners. Knowing this progression of skills helps teachers decide on the next best instructional step for students, as well as appreciate that all strategies are not of equal sophistication. Study the two student responses to the same equal groups problem in Figures 1.10 and 1.11. How are they similar and how are they different?

Figure 1.10 Philip's response. Philip wrote 6 eight times and then used a building up strategy to add.

There are 8 ants in an ant farm. Each ant has 6 legs. How many legs do the ants have altogether?

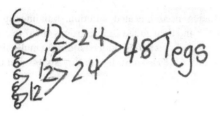

Figure 1.11 Jada's response. Jada uses the known fact of 8 × 5 to derive the fact 8 × 6.

There are 8 ants in an ant farm. Each ant has 6 legs. How many legs do the ants have altogether?

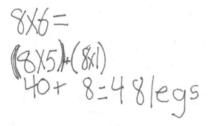

As you probably noticed both students arrive at the correct answer of 48 legs on 8 ants but use very different strategies. Philip's work shows evidence that he is still thinking about multiplication as repeated addition. Jada's work shows that she has an understanding of the distributive property and uses it to derive a math fact she does not know. These two strategies indicate very different understanding and call for different instructional responses. The examples are an illustration of the idea that all strategies are not equal in sophistication.

Throughout the book the *OGAP Multiplication and Division Progressions* will be used to help you understand evidence in student work and make instructional decisions.

All Strategies Are Not Equally Sophisticated

Take a minute to examine Braden's work in Figure 1.12. Braden's teacher gave him a pre-assessment before beginning a unit on multiplication and division. This is Braden's response to one of the questions on the pre-assessment. She noticed that although he got most of the questions correct in the pre-assessment, he used a strategy like the one shown in Figure 1.12 throughout.

Figure 1.12 Braden's response. Braden used repeated addition, building up from 14 groups of 437 to get 56 groups of 437 and then added 2 more groups of 437 to account for 58 groups. His solution is accurate and shows a solid understanding of multiplication as evidenced by his recognition that four 6118s is equivalent to fifty-six 14s, but is very inefficient.

Solve 437 × 58.

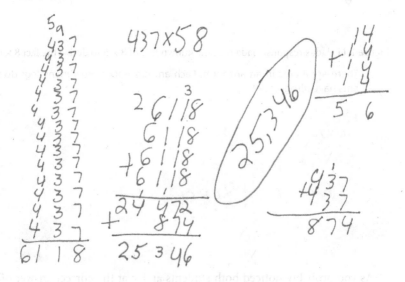

Braden's teacher wondered why a student would use addition when it was clear from his work that he had strong reasoning and an understanding of multiplication far beyond adding 437 together 57 times. She went back to his teacher from the year before to talk about Braden. His previous teacher was very surprised by this work and said that Braden had been using multiplicative strategies to solve problems consistently last year. The teacher then asked Braden why he used this strategy. Braden's response surprised the teacher. He said his strategy was based on the class discussion the day before, that multiplication was just repeated addition.

One lesson the teacher learned from this experience is the importance of gathering pre-assessment evidence of student understanding before introducing a topic at a level below or above a student's level of understanding. She also learned the importance of paying close attention to the strategies a student is using, not just whether the answer is correct.

Earlier in the chapter we discussed that our goal is for students to have a variety of strategies they understand and know when to use. In order to do this students need a deep understanding of the concepts to draw from that allow them to reason multiplicatively and use efficient and flexible strategies to solve problems, as was evidenced in Ethan's work in Figures 1.6 to 1.9. Although students can easily add 4 together 5 times to solve 4×5, it becomes a limiting strategy when solving multidigit multiplication problems such as 437×58. We want students to strive for efficiency and flexibility. We know there are multiple ways to solve a problem, but we cannot lose sight of the fact that not all strategies are equally efficient or sophisticated, and many strategies leave students unprepared for the multiplicative demands of more complex mathematics.

As teachers, we must push students toward more efficient methods that rely on multiplicative reasoning while assuring they preserve the less efficient strategies as part of their understanding so they can fall back on a wide range of strategies when needed. Explicitly linking strategies to each other and making connections clear to students is essential for this to occur. Ultimately our goal is for students to be strong multiplicative reasoners, regardless of the context or problem structures.

As you probably already appreciate, students, as well as teachers, benefit when they have an understanding of how the multiplicative skills and concepts work together. We would not consider a student to be a strong multiplicative reasoner if they only possessed one or two of these skills and concepts. Teachers can use the OGAP Multiplication and Division Progressions as a tool to help both themselves and students understand the progression of sophistication of strategies and the links between those strategies.

Consider the following vignette. The fifth grade teachers in this situation understand their students need both conceptual understanding and procedural fluency, and simply getting the correct answer is not enough.

Near the end of the third quarter of the school year, when all the units on multiplication and division had been taught, the 5th grade teachers in one school gave their students a post-assessment. Although the results on the post-assessment were good overall, the teachers noticed many of their students were accurately using repeated addition to solve most of the multiplication problems. These students were getting the correct answers, but their strategies were not at the sophistication level that is expected by the end of grade 5. They discussed the problem with their administrator and as a group decided to schedule an additional

30-minute block of time for the next month to focus on increasing the sophistication levels of their students' strategies. At the end of four weeks they gave another post-assessment. Most students had abandoned the repeated addition for more efficient multiplicative strategies. When thinking about what they would do in the future to avoid this situation, the teachers decided to be more purposeful in their use of the *OGAP Multiplication and Division Progressions* as the year progresses to assure that students develop strategies at the *Multiplicative* level.

The Importance of Multiplicative Reasoning

The Common Core State Standards for Mathematics

The development of multiplication and division understanding and fluency has a central role in the grades 3–5 CCSSM. In Chapter 2, and throughout the book, we will discuss specifically what the multiplication and division expectations are for grades 3–5 in the CCSSM (CCSSO, 2010). In addition, many concepts in grades 3–5 that are not specifically about teaching and learning multiplication and division rely on multiplicative reasoning. This includes aspects of fractions, decimals, geometry, and measurement.

Multiplicative Reasoning in Middle School

As Vergnaud (1983) indicated in the quote at the beginning of this chapter, multiplicative reasoning provides the foundation for much of the mathematics students interact with later, both in their school career and life. The importance of having robust multiplicative reasoning before entering middle school is critical and has been documented in studies of students' development of proportional reasoning (Hart et al., 1981; Lin, 1991; Tourniaire & Pulos, 1985). Because of this, by the time a student completes fifth grade, they must have both a strong foundation in multiplication and division concepts and a variety of strategies to use to approach both familiar and unfamiliar content.

In the K–3 mathematics curriculum, students build their additive reasoning. The mathematics in grades 3–5 focuses on building students' multiplicative reasoning. As students move into middle grades they encounter a number of topics in mathematics that rely heavily on multiplicative reasoning: fractions, rates, ratios, and proportions to name a few. Research has documented the inadequacy of additive reasoning for approaching these content strands in middle school (Sowder et al., 1998). Therefore, it is essential that students know how to reason multiplicatively and can distinguish situations that require multiplicative reasoning. This means that it is equally important for teachers to possess the knowledge of how students build and acquire multiplicative reasoning and what to do instructionally when students struggle. This book has been written to help build that understanding.

Chapter Summary

This chapter focused on:

- The difference between multiplicative and additive reasoning, considering both the number relationships and actions of each operation,
- The difference between relative and absolute reasoning,
- Characteristics of strong multiplicative reasoning,
- The importance of both conceptual understanding and procedural fluency in multiplicative reasoning,
- An introduction to the *OGAP Multiplication and Division Progressions*.

Looking Back

1. **Making Sense of the *OGAP Multiplication Progression*:** Both the *OGAP Multiplication and Division Progressions* communicate the progression of skills and concepts students should acquire as they become multiplicative reasoners. Examine the *OGAP Multiplicative Progression* and answer the following questions.

   (a) Based on the sample solutions shown on the progression, what are the key differences between *Additive* and *Early Transitional* strategies?

   (b) What are the most important differences between *Early Transitional* and *Transitional* strategies?

   (c) What are the major characteristics of *Multiplicative* strategies?

2. **What Is Multiplicative Reasoning?** Imagine you are speaking with parents of your third grade students at a September "Welcome Back to School" evening. You want to describe to them that the major emphasis of the mathematics in third grade is multiplicative reasoning, and this differs substantially from the earlier grades where additive reasoning was the focus. How would you describe multiplicative reasoning to these parents?

3. **Differentiating Additive Reasoning from Multiplicative Reasoning:** We learned in this chapter that even though students can use additive strategies to solve whole number multiplication problems, multiplication and division require more than simply extending additive concepts. Examine the array in Figure 1.13.

   Figure 1.13 Array of 12 circles.

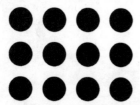

(a) List the different ways a student might utilize additive reasoning to determine the total number of circles in this array. Notice how the action of "joining" manifests itself in each example you provided.
(b) List the different ways a student might utilize multiplicative reasoning to determine the total number of circles in this array. Explain how each example you listed is an example of the "many to one" concept.

4. **Multiple Copies of a Unit:** An important aspect of multiplicative reasoning is the ability to count iterations of a unit. Review Philip's solution to the ant farm problem shown in Figure 1.10 earlier in the chapter. This solution is an example of "building up", an *Early Transitional* strategy grounded in repeated addition. As he is building toward the total of 48 legs, notice he notates several different quantities. We can think of each quantity as a unique unit that can be iterated. Identify what the following quantities in Philip's solution represent given the context of the problem.
(a) What does each "6" represent?
(b) What does each "12" represent?
(c) What does each "24" represent?

5. **Absolute and Relative Reasoning:** Although we use both absolute and relative thinking to compare quantities, students must be able to reason relatively in order to make sense of more complex mathematical concepts, such as those related to fractions, decimals, and proportions, to name a few. Consider the following situation. Identify whether each question that follows requires absolute or relative thinking? How do you know?

The Spring Concert

Two teams of students sold tickets to the Spring Concert. Timothy's team sold 24 tickets and Delia's team sold 36 tickets.

(a) How many more tickets did Delia's team sell than Timothy's team?
(b) How many times more tickets did Delia's team sell than Timothy's team?
(c) How many tickets did the two teams sell in all?
(d) What fraction of all the tickets sold did Timothy's team sell?

2

The OGAP Multiplication Progression

Big Ideas

- The *OGAP Multiplicative Framework* is based on math education research on how students develop multiplication and division fluency with understanding and is designed as a tool for teachers to gather evidence of student thinking to inform instruction and monitor student learning.
- Accumulating evidence by researchers indicates that knowledge and use of learning progressions positively affects both teachers' knowledge and instruction and students' motivation and achievement.

The *OGAP Multiplicative Framework* was developed from mathematics education research on how students learn multiplicative reasoning concepts and is a valuable tool to help teachers select or design tasks, understand evidence in student work, make instructional decisions, and provide actionable feedback to students. This chapter provides an overview of the *OGAP Multiplication Progression*, which is one aspect of the *OGAP Multiplicative Framework*. The *OGAP Division Progression* is discussed in Chapter 7.

It is suggested that you download the *OGAP Multiplicative Framework* and refer to it as you read this chapter and references to the framework in other chapters of the book. An electronic copy can be found at www.routledge.com/9781138205697.

There are two major elements in the *OGAP Multiplicative Framework*:

1. *Problem Contexts and Structures* (front page) and *Sample Problems* (back page)
2. The *OGAP Multiplication and Division Progressions* that show evidence of student work along a continuum of student understanding for multiplication and division (pages 2–3)

The parts of the framework are interrelated. That is, movement on the progression is often influenced by the structures of the problems as students are developing their multiplicative reasoning and fluency. Student work samples are used throughout this chapter to describe and exemplify the different levels on the progression. Strategies to help move student understanding along the progressions are embedded in this and other chapters throughout the book. *Problem Contexts and Structures* and the *OGAP Division Progression* are fully discussed in Chapters 5, 6, and 7, respectively.

The *OGAP Multiplication Progression*

The *OGAP Multiplication Progression* is designed to help teachers gather descriptive evidence of student thinking related to developing understanding of multiplicative reasoning concepts and skills, as well as to identify the underlying issues and errors that may interfere with students learning new concepts or solving multiplication and division problems. The *OGAP Multiplication Progression* also provides some instructional guidance on how to transition student understanding and strategies from one level to the next, with the goal of developing procedural fluency. Researchers indicate that students may struggle with the use and understanding of formal algorithms if their knowledge is dependent on memory, rather than anchored with a deep understanding of the foundational concepts (e.g., Battista, 2012; Carpenter, Franke, & Levi, 2003; Empson & Levi, 2011; Fosnot & Dolk, 2001; Kaput, 1989). The importance of developing fluency in a way that brings meaning to both cannot be overstated.

Additionally, there is accumulating evidence that knowledge and instructional use of learning progressions, together with the mathematics education research that underpins progressions, positively affects instructional decision-making and student motivation and achievement in mathematics (Carpenter, Fennema, Peterson, Chiang, & Loef, 1989; Clarke, 2004; Clements, Sarama, Spitler, Lange, & Wolfe, 2011; Fennema, Carpenter, Levi, Jacobs, & Empson, 1996). This research supports the use of a progression as an effective strategy to gather and act on evidence of student thinking as students develop understanding and fleuncy with multiplication and division.

When you review the *OGAP Multiplication Progression* you will notice that the levels reflect different kinds of evidence that might be found in student work as students learn new concepts and solve problems. Each level on the progression is briefly described in this section. However, as you work through the other chapters in this book, there will be opportunities to deepen your understanding of the progression.

OGAP Multiplication Progression Levels

The *OGAP Multiplication Progression* levels represent the continuum of evidence from *Nonmultiplicative* to *Multiplicative Strategies* that is visible in student work as students develop their understanding and fluency with whole number multiplication and division. The levels are at a grain size that is usable by teachers to gather actionable evidence across the development of multiplication concepts and skills.

Open to pages 2 and 3 of the *OGAP Multiplicative Framework*. You will notice the *OGAP Multiplication Progression* on page 2 and the *OGAP Division Progression* on page 3. As indicated earlier, this chapter focuses on the *OGAP Multiplication Progression*. Notice the six levels along the left side of the progression: *Nonmultiplicative Strategies, Early Additive Strategies, Additive Strategies, Early Transitional Strategies, Transitional Strategies,* and *Multiplicative Strategies*.

The progressions are designed to provide an expected path based on mathematics education research that supports the development of procedural fluency with understanding. The *Transitional* level (including *Early Transitional*) of the progression is the bridge between additive strategies and multiplicative strategies.

Early Additive Strategies

When students first engage in equal groups multiplication and division problems, they often draw on their counting skills to solve the problems.

Samantha's response in Figure 2.1 shows evidence of understanding the problem situation by sketching each tricycle and then counting each wheel. This is evidenced by tic marks in each wheel, which suggests that Samantha touched each wheel with her pencil and counted. This strategy is evidence of *Early Additive* thinking because by modeling the situation, she has turned it into a counting problem, in this case counting by ones.

Figure 2.1 *Early Additive Strategy*. Samantha's response illustrates evidence of a counting by ones strategy to solve this equal groups multiplication problem.

How many wheels do 29 tricycles have? Show your work.

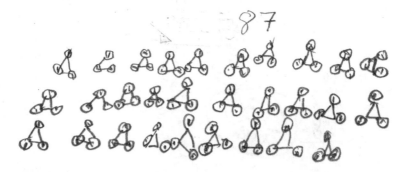

From Early Additive to Additive

As students move away from *Early Additive Strategies* to *Additive Strategies* they begin to unitize and operate with equal groups—for example, by iterating the composite units and then applying repeated addition as evidenced in Tammy's and Hunter's solutions in Figure 2.2 and 2.3, respectively.

Figure 2.2 *Additive Strategy.* Tammy's response shows evidence of unitizing into equal groups of 3 and adding the groups in this array.

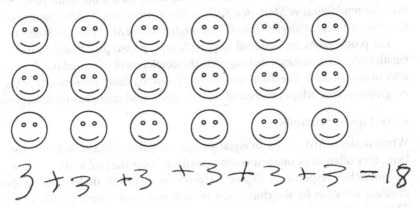

$$3 + 3 + 3 + 3 + 3 + 3 = 18$$

Figure 2.3 *Additive Strategy.* Hunter's accurate solution shows evidence of coordinating the composite unit (e.g., 35 lbs. per bag) and the number of bags (e.g., 9 bags) using a repeated addition strategy.

Max and Thomas each delivered vegetables to a store. Max delivered 8 bags of vegetables with 40 pounds in each bag. Thomas delivered 9 bags of vegetables with 35 pounds in each bag. How many pounds of vegetables were delivered altogether?

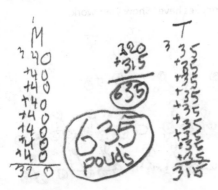

Notice the significant differences between Samantha's strategy and the strategies in Tammy's and Hunter's solutions. Rather than counting each individual object as Samantha did, Tammy and Hunter counted groups of objects. Counting by groups shows evidence of unitizing, or the ability to conceptualize a group of individual things as one group, and is an important step to developing multiplicative reasoning (Fosnot & Dolk, 2001).

 Go To For an in-depth discussion about unitizing go to Chapter 4: The Role of Concepts and Properties.

From Additive to Early Transitional

Hunter's repeated addition strategy in Figure 2.3 is accurate and shows evidence of unitizing. However, it is not efficient, nor does it show evidence of understanding the multiplicative relationships so important to developing procedural fluency with understanding. The focus at the *Transitional* level is to *bridge* students from additive strategies and reasoning (e.g., counting by ones, counting by equal groups) to procedural fluency with understanding at the multiplicative level.

At the *Early Transitional Level*, building up strategies (Figure 2.4) and skip counting (Figure 2.5) are evidenced as students begin to combine groups.

Figure 2.4 *Early Transitional Strategy.* Owen's correct response shows evidence of building up by combining groups.

What is the area of a closet that is 5 feet by 6 feet? Show your work.

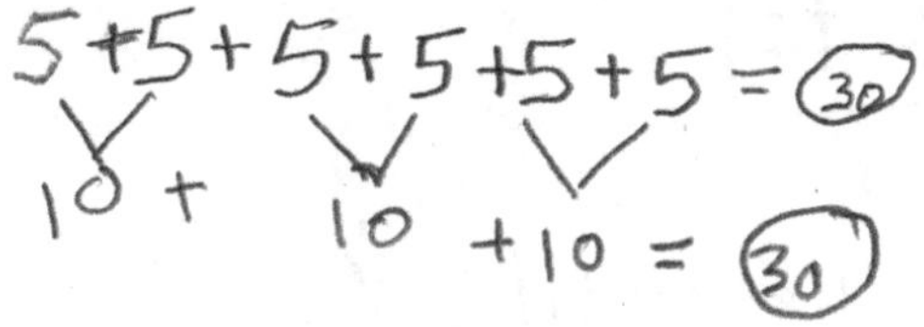

$$5+5+5+5+5+5 = 30$$
$$10 + 10 + 10 = 30$$

Figure 2.5 *Early Transitional Strategy.* Jack's correct solution shows evidence of using skip counting.

Max and Thomas each delivered vegetables to a store. Max delivered 8 bags of vegetables with 40 pounds in each bag. Thomas delivered 9 bags of vegetables with 35 pounds in each bag. How many pounds of vegetables were delivered altogether?

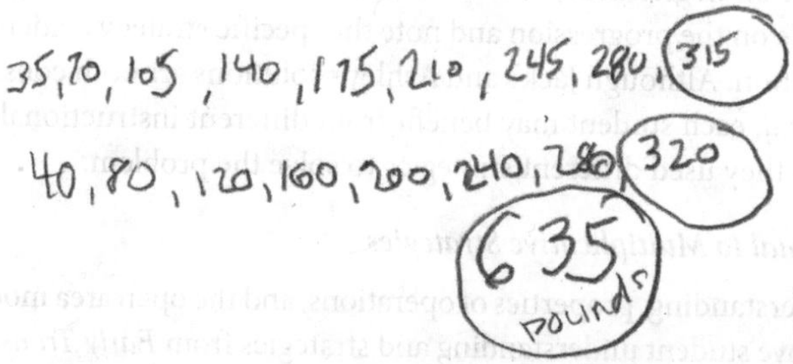

35, 70, 105, 140, 175, 210, 245, 280, 315

40, 80, 120, 160, 200, 240, 280, 320

635 POUNDS

For example, in Figure 2.4 Owen is able to combine 6 groups of 5 into 3 groups of 10 to find the total of 20. In Jack's solution in Figure 2.5, his skip counting reflects 35 as 1 group of 35 pounds, 70 as 2 groups of 35 pounds, 105 as 3 groups of 35 pounds, and so on.

As illustrated in this section, at the *Early Transitional* level students are still operating with groups, but combining groups using strategies such as skip counting and building up. Key to moving from these *Early Transitional Strategies* toward more sophisticated *Transitional Strategies* is the use of the area model. The area model can be used to help students transition from seeing single units, to groups of units in rows or columns, to ultimately understanding the multiplicative relationship between the two dimensions of a rectangle. Ashley's solution in Figure 2.6 shows evidence of considering both dimensions in the rectangle that she drew; an important step toward thinking multiplicatively.

Figure 2.6 *Early Transitional Strategy.* Ashley used an area model to solve the problem.

What is the area of a closet that is 5 feet by 6 feet? Show your work.

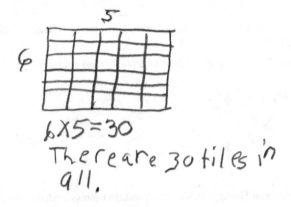

Although the solutions in Figures 2.5 and 2.6 are both classified as *Early Transitional Strategies* they represent a very different level of understanding. This points to an important aspect of using the *OGAP Multiplication Progression*. In order to make appropriate instructional decisions based on evidence of student thinking or strategies used, one must consider more than the level of the student response on the progression and note the specific strategy evidenced in the student solution. Although Jack's and Ashley's solutions are both considered *Early Transitional*, each student may benefit from different instructional experiences because they used different strategies to solve the problem.

From Transitional to Multiplicative Strategies

Place value understanding, properties of operations, and the open area model are used to help move student understanding and strategies from *Early Transitional*

Strategies to *Transitional* to *Multiplicative Strategies.* In Figure 2.7 Hank decomposed the factors (16 and 24) using place value understanding. He then used the open area model to solve the problem. The open area model used in this way is a direct link to the distributive property and the partial products and traditional algorithms for multiplication. The open area model also helps students consider the impact of place value in multiplication in ways that are masked by the traditional algorithm. In the following example, you can see that the product of 20×10 is much larger than the product of 4×6.

Go To For discussions about the use of properties of operations and the open area model to support the development of fluency with understanding go to Chapter 3: The Role of Visual Models; Chapter 4: The Role of Concepts and Properties; Chapter 8: Understanding Algorithms; and Chapter 9: Developing Math Fact Fluency.

Figure 2.7 *Transitional Strategy.* Hank effectively used an open area model to solve the problem.

There are 16 players on a soccer team. How many players are in a league if there are 24 soccer teams? Show your work.

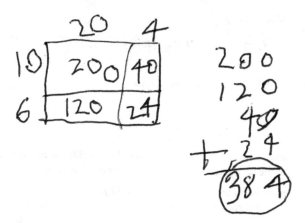

Multiplicative Strategies

At this level there is evidence that students use efficient and flexible strategies (e.g., partial products, traditional US algorithm, distributive property), as well as appropriately apply properties of operations (e.g., associative, commutative, and distributive) to solve multiplication and division problems. At this stage, students no longer need to use models to support their thinking about multiplication and division. By the end of fifth grade and beginning of sixth grade students should be using *Multiplicative* strategies to solve whole number problems involving equal groups, equal measures, measurement conversions, patterns, multiplicative comparisons, unit rates, rectangular area, and volume problems.

Go To For an in-depth discussion of the different contexts that students interact with as they develop their multiplication and division fluency and understanding, go to Chapter 5: Problem Contexts.

Figures 2.8 to 2.10 are examples of student solutions at the *Multiplicative* level. What understandings are evidenced in each response? How are they alike, and how are they different?

Figure 2.8 *Multiplicative Strategy.* Kylee's response shows evidence of flexibly using place value understanding and the distributive property.

> Max and Thomas each delivered vegetables to a store. Max delivered 8 bags of vegetables with 40 pounds in each bag. Thomas delivered 9 bags of vegetables with 35 pounds in each bag. How many pounds of vegetables were delivered altogether?

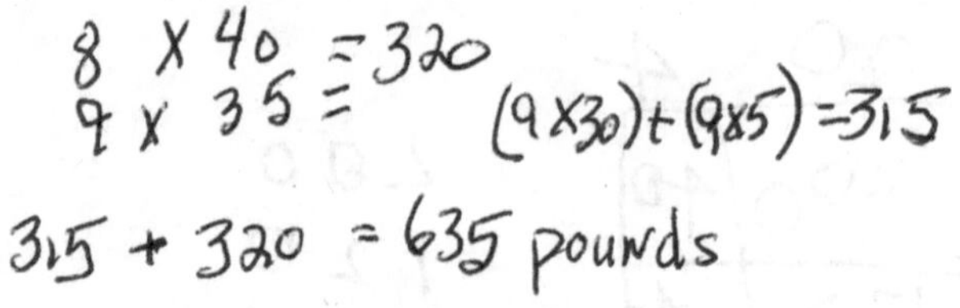

$$8 \times 40 = 320$$
$$9 \times 35 = \quad (9 \times 30) + (9 \times 5) = 315$$
$$315 + 320 = 635 \text{ pounds}$$

Figure 2.9 *Multiplicative Strategy.* Michael's response shows evidence of using place value understanding in the accurate use of the partial products algorithm.

> A family is carpeting a room with the shape and dimensions pictured. How many square feet of carpet will they need to cover the whole floor?

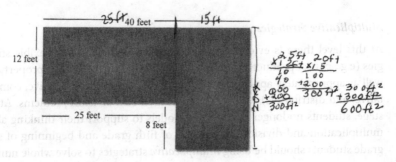

Figure 2.10 *Multiplicative Strategy.* Thomas's response shows evidence of correctly using the US traditional algorithm.

There are 16 players on each team in the Smithville Soccer League. How many players are in the league if there are 24 teams?

$$
\begin{array}{r}
{}^{1}\,{}^{2} \\
16 \\
\times\ 24 \\
\hline
64 \\
3\,20 \\
\hline
384
\end{array}
$$

384 players

Although the strategies evidenced in the student work for each of these responses is at the multiplicative level, you probably noticed that they show different understandings. That evidence can be used to inform the next instructional steps for each of these students. For example, Kylee flexibly used place value understanding when multiplying 8×40, when decomposing 35 into $30 + 5$, and then accurately applied the distributive property to solve the problem. Because the problem involved multiplying a single-digit number by a two-digit number, one of which was a multiple of a power of 10 (40 pounds), you may want to collect additional evidence about how Kylee would approach solving a problem that involves multiplication of a two-digit by a two-digit number.

The idea presented in this example is important. That is, even though Kylee accurately solved the problem at the *Multiplicative Strategy* level, the evidence should be used in the same way as evidence at the other levels of the progression to inform next instructional steps. In the example earlier we suggested changing the magnitude of the factors to gather additional evidence about the strategy that Kylee would use when the factors were more complex, but we could have asked Kylee a question that involves a more complex problem context (e.g., multiplicative comparison, area) as well. The point is, as you begin to look at the evidence through the lens of the *OGAP Multiplication and Division Progressions*, think similarly about using the evidence to inform instruction at the *Multiplicative* level as you would for student work at the other levels of the progression.

Nonmultiplicative Strategies

Notice the section labeled *Nonmultiplicative Strategies* at the bottom of the *OGAP Multiplication Progression*. As students engage in new topics or are just beginning multiplication concepts, they often add factors, use the incorrect operation, or guess. The solution in Figure 2.11 shows evidence of adding the factors.

What is interesting and important to note, and more fully described through examples in Chapter 6: Problem Structures, is that students may be using multiplicative strategies for one problem context (e.g., equal groups) and then revert to a nonmultiplicative strategy when first solving a new problem context (Figure 2.12).

Figure 2.11 *Nonmultiplicative Strategy.* Madison's response shows evidence of adding the factors instead of multiplying 23 inches of string times the number of decorations.

Twenty-three inches of string are needed to hang each decoration. How many inches are needed to hang 9 decorations?

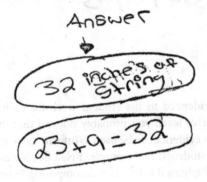

Figure 2.12 *Nonmultiplicative Strategy.* Robert used division instead of multiplication in part B.

One tricycle has three wheels.

(a) How many wheels do 5 tricycles have?

$$5 \times 3 = 15$$

(b) How many wheels do 29 tricycles have?

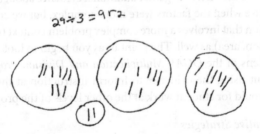

Robert's response in Figure 2.12 illustrates instability that is often evidenced as students are developing initial understanding. Although problems A and B are identical except for the magnitude of the numbers in the problem (A. 3 × 5;

B. 3 × 29), the student solves Part A as a multiplication problem and Part B as a division problem. Because the wrong operation was used to solve B, the solution is classified as *Nonmultiplicative*. There are many reasons why the student may have solved Part B as division, but the decision probably was based on the magnitude of the number (29 tricycles), not a misunderstanding of the problem situation. Importantly, the best way to really understand what influenced the student's solution in Part B is to ask the student.

Important Ideas about the *OGAP Multiplication and Division Progressions*

This section focuses on the following important points to keep in mind when using both the *OGAP Multiplication and Division Progressions*:

1. Movement along the progressions is not linear.
2. Students' strategies will be at different levels on the progression at different times.
3. The progressions provide instructional guidance.
4. The progressions are not evaluative.
5. Collection of *Underlying Issues and Errors* is important.

1. **Movement along the *OGAP Multiplication and Division Progressions* is not linear.** Although the progression looks linear, student development of understanding and fluency is a more complex pathway. As students are introduced to new concepts, different problem structures for the same concept, more complex numbers, or asked to apply their multiplicative reasoning knowledge to other mathematical topics, their solution strategies may move back and forth between multiplicative, transitional, additive, and nonmultiplicative strategies and reasoning based on the strength of their multiplicative reasoning.

2. **Students' strategies will be at different levels** on the progression depending on the concepts being taught and learned or the problems they are solving. That is, a student may be using a *Multiplicative Strategy* when solving an equal groups problem and a *Nonmultiplicative Strategy* when solving an area problem. Figure 2.12 (tricycles) is an example of a solution with evidence at two levels: *Multiplicative* (known fact) and *Nonmultiplicative* (wrong operation).

The graphic in Figure 2.13 illustrates these important points. That is, as multiplicative reasoning, understanding, and fluency develop, and as students are introduced to new concepts or interact with different problem structures for the same concept, students' solutions may move up and down the progression levels (Kouba & Franklin, 1993; OGAP, 2006). On the left side of the *OGAP Multiplication Progression*, the two-way arrow represents this important idea.

By middle school multiplicative reasoning, fluency, and understanding should be stabilized at the *Multiplicative* level for whole number multiplication and division so that students can fully engage in middle school mathematics.

Figure 2.13 Hypothesized movement on the progression as concepts are introduced and developed across grades.

Source: Adapted from (Petit, Laird, Marsden, & Ebby, 2015).

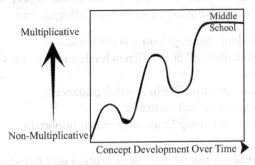

3. **The *OGAP Multiplication and Division Progressions* provide instructional guidance.** Along the right side of both the *OGAP Multiplication and Division Progressions* there is an arrow that contains key concepts (unitizing, uses visual models, understanding of place value, and properties of operations) important for the development of multiplicative fluency with understanding. These concepts serve as instructional guidance on how to move student understanding and strategies from one level to the next. For example, if a student is still solving multiplication problems counting by ones, then a focus on developing unitizing is important. On the other hand, if a student consistently uses skip counting, then a focus on using visual models transitioning from equal groups to area models may be called for.

Go To For an in-depth discussion about these concepts and their role in the development of multiplication and division understanding and fluency, go to Chapter 3: The Role of Visual Models and Chapter 4: The Role of Concepts and Properties.

There are a couple of important ideas to consider as you make decisions about the next instructional step based on evidence of student thinking and strategies when using the *OGAP Multiplication Progression* that can be best understood through an example. Consider Samantha's solution in Figure 2.1 in which she drew each tricycle and counted each wheel. The next instructional step for Samantha is not teaching her a formal algorithm because she probably does not yet have the foundation of multiplicative understanding. On the other hand, the next instructional step would not be 'modeling and counting by subgroups' which is another *Early Additive Strategy*. Rather, instruction should focus on unitizing or conceptualizing equal groups using strategies such as subitizing (see Chapter 3) and then counting by groups. This example illustrates that one should not skip past the important transitional stages that are designed to build fluency with conceptual understanding. At the same time, one does not need to engage students in every strategy at every level if the strategy does not support the forward movement of student understanding.

Using the *OGAP Multiplication Progression* to guide instruction is not about direct instruction on specific strategies or concepts. Rather, it involves the interaction of foundational concepts (e.g., unitizing, place value, and properties of operations) with targeted instructional strategies (e.g., connecting mathematical ideas, classroom discourse, sharing solutions, and purposeful questioning) to engage students in thinking and reasoning about these concepts and relationships.

4. **The *OGAP Multiplication and Division Progressions* are not evaluative.** You'll notice that there are no numbers associated with the levels on the progressions. A learning progression is designed to help teachers gather descriptive evidence about student learning to inform instruction and student learning, not to assign a number or grade. The descriptive evidence includes the level on the *OGAP Multiplication and Division Progressions*, the substrategy, underlying issues or errors, and evidence of solution accuracy (e.g., *Early Transitional*, skip counting, calculation error, incorrect).

5. **Collection of *Underlying Issues and Errors* is important.** At the bottom of the progressions there is a list of potential underlying issues or errors that may interfere with students learning new concepts and solving related problems. Sometimes evidence of errors or underlying issues do not influence where the evidence is classified along the progression. For example, consider a student who uses an open area model to solve a problem at the *Transitional Level* but makes a calculation error. It is important to record this calculation error because it affects accuracy, but it does not change the level on the progression. Other times an error can influence the placement on the progression—errors like using the wrong operation, guessing, or adding factors (Figure 2.11). This information, coupled with the location of the strategy used along the progression, provides teachers with actionable evidence to inform instruction and learning.

The CCSSM and the *OGAP Multiplication and Division Progressions*

The CCSSM reflects the progression of developing procedural fluency with understanding using visual models (e.g., arrays, area model, open area model), place value, and properties of operations. As you study Table 2.1, you will notice a pattern in the CCSSM similar to the *OGAP Multiplication and Division Progressions*. Students first engage with multiplication at grade 2 through repeated addition of equal groups in an array to develop a foundational understanding of multiplication. Then students use place value, properties of operations, and visual models to develop fluency with understanding. Students should show evidence of consistently solving a range of problems using efficient algorithms for whole number multiplication by grade 5 and division at grade 6

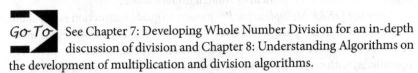

Go To See Chapter 7: Developing Whole Number Division for an in-depth discussion of division and Chapter 8: Understanding Algorithms on the development of multiplication and division algorithms.

It is important to note that the CCSSM reflects research that students should not be taught efficient algorithms prematurely (e.g., Ebby, 2005; Kamii, 1998; Wong & Evans, 2007). Rather, students should develop whole number multiplication and division fluency with understanding in grades 2 through 4. As explained earlier, the goal for all students is fluency by grade 5 for multiplication and grade 6 for division that is built upon strong understanding developed in the earlier grades.

Table 2.1 CCSSM expectations for multiplicative strategies.

Grade	CCSSM Expectations for Multiplicative Strategies (CCSSO, 2010)
2	Use repeated addition to find the total of equal groups in an array
3	Solve multiplication and division problems using strategies based on place value and properties of operations (e.g., commutative, associative)
4	Solve multiplication and division problems using strategies based on place value (e.g., partial products, area models) and the properties of operations and relationships (e.g., commutative, associative, and distributive; inverse relationship between multiplication and division)
5	Solve multiplication problems using a standard algorithm (e.g., partial products algorithm, traditional US algorithm for multiplication)
	Solve division problems using strategies based on place value (e.g., partial quotients, menus, area models) and the properties of operations and relationships (e.g., the inverse relationship between multiplication and division)
6	Solve division problems using standard algorithms

Using the *OGAP Multiplication and Division Progressions* as an Instructional Tool

As has been mentioned earlier in this chapter and throughout the book, the *OGAP Multiplicative Framework* was developed as part of a formative assessment project. The OGAP process is cyclical. It involves eliciting evidence of student thinking and adjusting instruction as students are learning on an ongoing basis: during classroom discussions, observations as students are working, and collecting evidence from each student at the end of lessons.

Strategies to systematically collect evidence to inform instruction and student learning are an essential part of OGAP and were developed and refined through interactions with hundreds of teachers over the last decade. One strategy OGAP teachers have adapted is the regular use of *exit questions* at the end of a lesson. Although teachers should be observing, listening, and adjusting instruction as lessons progress, an exit question at the end of a lesson provides evidence from each student that can inform instruction for the next lesson.

When teachers analyze evidence from exit questions, they use the *OGAP Sort*. *The OGAP Sort*, more fully described later, provides evidence to inform instruction with regard to three aspects of student solutions:

1. Level on the progression with detailed strategy evidenced
2. Underlying issues or errors
3. Accuracy of solution

Traditionally, when looking at student work, the accuracy of the solution is the first filter. However, OGAP studies have shown that accuracy alone may produce a "false positive." For example, a fifth grade student who consistently uses repeated addition to solve multiplication problems will be at a significant disadvantage when engaging in middle school topics dependent upon strong multiplicative reasoning and fluency (e.g., proportions). That is, a correct answer may hide the fact that the student may not have the needed multiplicative understanding or strategies to engage in middle school math topics. Recall Hunter's work in Figure 2.3 where he added up 40 eight times and 35 nine times to find the solution. Although he correctly answered the question, this apparent level of understanding may interfere with more sophisticated concepts in subsequent years.

To focus the analysis of evidence from a progression perspective, the *OGAP Sort* begins with organizing the student work into piles that correspond with each of the progression levels as shown in Figure 2.14.

Once sorted the information is then recorded on an *OGAP Evidence Collection Sheet* such as the one pictured in Figure 2.15. Next, one returns to the student work and makes notes about the strategies used within a level and any *Underlying Issues or Errors*. In the example in Figure 2.15 the teacher also highlighted the solutions that were incorrect.

Figure 2.14 Sorting student work into *Multiplicative, Transitional, Additive,* and *Non-multiplicative* strategies.

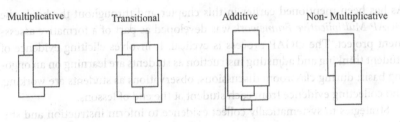

Multiplicative   Transitional   Additive   Non-Multiplicative

This collection sheet parallels the *OGAP Multiplicative Reasoning Progression* and provides teachers with a picture of where the evidence in the student solutions is along the progression, the errors that may be interfering with learning, and the accuracy of the answer. Together, these pieces of information can help inform instruction for the whole class, for small groups, and for individual students.

Figure 2.15 Sample of completed *OGAP Student Work Evidence Collection Sheet.* Student names are listed under strategy level evidenced, substrategies are noted, student names are listed under *Underlying Issues or Errors* if evidenced, and incorrect solutions are highlighted.

Evidence Collection Sheet

Items #	Content (e.g., context, type of number)	Multiplicative	Transitional		Additive		Non-multiplicative Reasoning
			Transitional	Early	Additive	Early	
1	Equal group 8×12	Open area model	Ethan Natalie Grace Lucas Abdi Nathan	Emma/Skip Claire/Counting Kelyn - Area Model	Tyler Jacob Sophie Alexis	Logan Eli - Counting by ones	Charlotte - Added factor 20 used procedure incorrectly

Underlying issues or concerns								
Unreasonable	Misinterpret meaning of remainders	Place value error	Units inconsistent or absent	Property or relationship error	Calculation error	Equation error	Model error	Vocabulary error
	Abdi				Ethan Grace			

Once the evidence is recorded, the analysis turns to looking across the whole class using the following questions to help make instructional decisions:

1. What are developing understandings that can be built upon?
2. What issues or concerns are evidenced in student work?
3. What are potential next instructional steps for the whole class, for small groups, and for individuals?

The OGAP process is cyclical. After you have reviewed the work and made instructional decisions, you should again consider the structures in the problems students will solve during the next lesson and in the exit question. Once more, analysis of the exit question from this lesson can inform your planning for the next lesson.

Most chapters of *A Focus on Multiplication and Division: Bringing Research to the Classroom* include a section on the *OGAP Multiplication and/or Division Progressions* describing what the evidence in the student work for the topic under discussion would look like at different levels of the progression. We also suggest you analyze the student work examples throughout the book through the lens of the *OGAP Multiplication and Division Progressions.*

 The icon to the left is used throughout the book to indicate where the *OGAP Multiplicative Reasoning Framework or the Multiplication or Division Progressions* are discussed.

Chapter Summary

This chapter focused on learning progressions and specifically the *OGAP Multiplication Progression*. Many of the points made about the OGAP Multiplication Progression hold true for the *OGAP Division Progression* more fully discussed in Chapter 7.

- The *OGAP Multiplicative Framework* consists of two sections: *Problem Contexts and Structures* and the *OGAP Multiplication and Division Progressions*.
- The *OGAP Multiplication and Division Progressions* are examples of learning progressions founded on mathematics education research, written at a grain size that is usable across a range of multiplicative concepts and by teachers and students in a classroom.
- The *OGAP Multiplication and Division Progressions* were specifically designed to inform instruction and monitor student learning from a formative assessment perspective.
- The *OGAP Multiplication and Division Progressions* illustrate how student strategies progress from *Additive* to *Transitional* to *Multiplicative* strategies as they develop understanding of foundational concepts such as unitizing, place value, and properties of operations and as they encounter a range of multiplicative problem situations and structures.
- Analyzing evidence in student work using the OGAP *Multiplication and Division Progressions* provides important information about where students are in their understanding of concepts and use of multiplicative strategies.

Looking Back

1. **Become familiar with the *OGAP Multiplicative Framework*:** The *OGAP Multiplicative Framework* is composed of two sections: *Problem Contexts*

and *Problem Structures* on page 1 with sample problems on page 4, and the *Multiplication and Division Progressions* on pages 2 and 3.

(a) Examine these two sections of the *OGAP Multiplicative Framework* and answer the following two questions:

- What is the main purpose of each section of the *OGAP Multiplicative Framework*?

- In what ways is the information in the *OGAP Multiplicative Framework* important to teachers who teach multiplication and division?

2. **Use the *OGAP Multiplicative Progression* to sort student work samples:** The OGAP Sort refers to the task of understanding a student solution and determining the level on the *OGAP Multiplicative Progression* that the solution best matches. Sorting student work is the first step in using evidence in student solutions to inform instructional decisions. Sorting student work is described in the chapter section Using the OGAP Multiplication and Division Progressions as Instructional Tools as an Instructional Tool.

The tricycle problem and four student solutions are shown next. Determine the level of the *OGAP Multiplication Progression* that best matches each solution. Record the evidence that supports your decision.

The Tricycle Problem

One tricycle has three wheels. How many wheels do 5 tricycles have?

Figure 2.16 Benjamin's response to the tricycle problem.

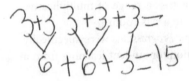

Figure 2.17 Ella's response to the tricycle problem.

15    $3 \times 5 = 15$
      $5 \times 3 = 15$

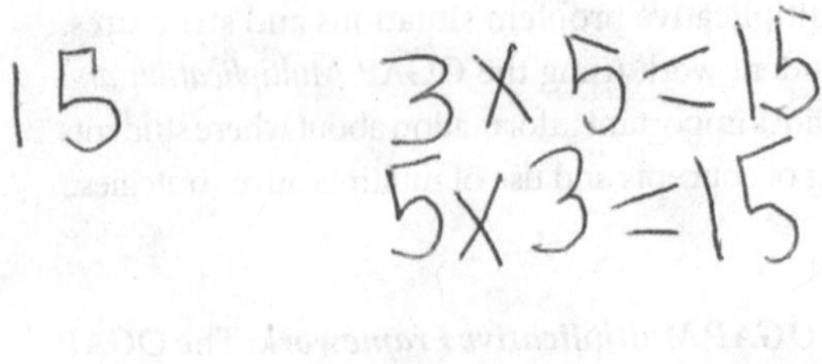

Figure 2.18 Rosie's response to the tricycle problem.

Figure 2.19 Glenn's response to the tricycle problem.

3. **Practice using student work to inform instruction:** Mr. Lorenzo used the following multistep problem to gather evidence of the ways four of his students are conceptualizing the important multiplication concepts they have been working on this past week.

The Photo Album Problem

The fourth grade class made an album of their favorite photos from across the school year. The album included 19 pages of small photos with 3 small photos per page. The album also included 23 pages of large photos with 2 large photos per page. What is the total number of photos in the album? Show your work.

Use the following student work and the *OGAP Multiplication Progression* to analyze the evidence in each of the solutions. Record your analysis on a copy of the *OGAP Evidence Collection Sheet* shown in Figure 2.24.
(a) For each solution identify:
  • The level on the progression the evidence is found. What is the evidence?
  • Any *Underlying Issues or Errors.*
  • Accuracy of the solution. Highlight the student solutions that are incorrect.

(b) Based on the evidence you collected, how might Mr. Lorenzo adjust his instruction in the next several lessons to best meet the needs of these four students?

Figure 2.20 John's response.

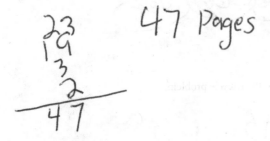

Figure 2.21 Ben's response.

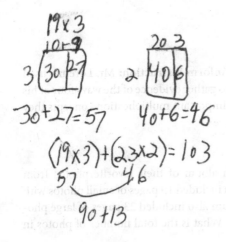

Figure 2.22 Emma's response.

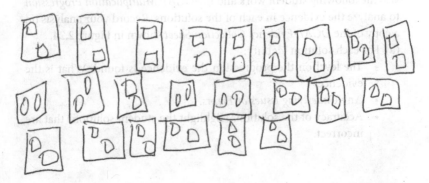

Figure 2.23 Hannah's response.

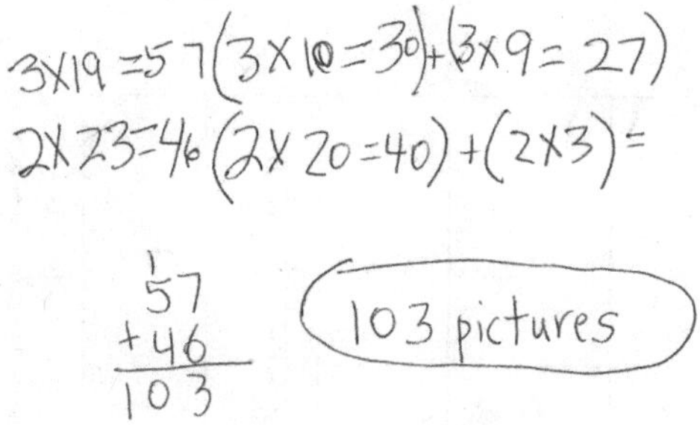

$$3 \times 19 = 57 \left(3 \times 10 = 30\right) + \left(3 \times 9 = 27\right)$$
$$2 \times 23 = 46 \left(2 \times 20 = 40\right) + \left(2 \times 3\right) =$$

$$\begin{array}{r} \overset{1}{57} \\ +46 \\ \hline 103 \end{array}$$ 103 pictures

4. **Try This in Your Class:** In Question 3, you practiced analyzing student solutions using the *OGAP Multiplication Progression* and the *OGAP Evidence Collection Sheet*. Most importantly, you considered instructional implications in light of the evidence you collected.

(a) Try this process with your students. Follow the three steps here, and use ideas from *Using the OGAP Multiplication and Division Progressions as an Instructional Tool* to guide this analysis.

   • Design or select a multiplication question based on the mathematical goal of the upcoming lesson.

   • Administer the question as an "exit question" at the end of the lesson.

   • Analyze your students' responses and record the information on a copy of the *OGAP Evidence Collection Sheet* shown in Figure 2.24

(b) Use the evidence you collected in 4a to answer the following three questions:

   • What are some developing understandings you noticed in the solutions that can be built upon in future lessons?

   • What are some underlying issues or concerns across your class that future lessons should address?

   • What are some implications for instruction, specific instructional actions you can take to address the evidence you collected?

Figure 2.24 OGAP Student Work Evidence Collection Sheet.

Evidence Collection Sheet

Items #	Content (e.g., context, type of number)	Multiplicative		Transitional		Additive		Non-multiplicative Reasoning
		Multiplicative	Transitional	Transitional	Early	Additive	Early	

Underlying Issues or concerns						
Unreasonable	Misinterpret meaning of remainders	Place value error	Units inconsistent or absent	Property or relationship error	Calculation error	Equation error	Model error	Vocabulary error

Instructional Link

Use the following questions to analyze ways your math instruction and program provide students opportunities to build fluency and understanding of important grade-level multiplicative concepts and skills.

1. To what degree do your math instruction and program focus on regularly gathering descriptive information about student learning to inform your instruction?
2. What are the ways in which your multiplication instruction uses strategies and tools such as unitizing, area models and arrays, and skip counting to help students transition from *Additive* to *Transitional* strategies?
3. How do you and your math program use place value, open area models, and properties of operations to develop understanding and fluency with vital grade-level multiplicative concepts and skills?
4. Based on this analysis, identify specific ways you can enhance your math instruction by utilizing ideas from this chapter.

3

The Role of Visual Models

Big Ideas

- Visual models play a key role in students' development of multiplicative reasoning and fluency.
- Use of quick images supports both subitizing and visualization and can help students transition to more sophisticated visual models and to mental models.

This chapter introduces the role of visual models in the development of fluency with understanding. Chapter 4 centers on unitizing, place value, and properties of operations. Importantly, both chapters provide research-based instructional strategies that support the development of understanding and fluency when multiplying and dividing whole numbers. The ideas related to the use of visual models, as well as the concepts presented in Chapter 4, will be expanded upon in Chapter 7: Developing Whole Number Division, Chapter 8: Understanding Algorithms, and Chapter 9: Developing Math Fact Fluency.

As you read Chapters 3 and 4 reflect on your own experiences learning to multiply and divide whole numbers. You may remember memorizing steps to complete calculations and practicing those steps over and over again. In contrast, researchers indicate that fluency and understanding should be built using visual models and the mathematical concepts underpinning the operations rather than just rote memorization of algorithmic steps (e.g., Battista, 2012; Carpenter et al., 2003; Empson & Levi, 2011; Fosnot & Dolk, 2001; Kaput, 1989).

As we begin this discussion, it is important to re-emphasize the point that procedural fluency is more than just being able to successfully use an efficient procedure to multiply and divide whole numbers. Rather, it means that students recognize a multiplicative relationship in a range of problem situations (e.g., equal groups, measure conversions, multiplicative comparisons) and find an efficient method to solve multiplication and division problems, regardless of the context, the magnitude or complexity of the numbers, or the number relationships.

The Role of Visual Models

Look closely at the different levels of both the *OGAP Multiplication and Division Progressions* on pages 2 and 3 of the *OGAP Multiplicative Reasoning Framework*. Notice the prominent role of visual models at different levels of the progression. There are two important perspectives to consider when using the progressions in relation to visual models: 1) understanding evidence of the sophistication of student strategies, and 2) guiding instructional decisions. Both perspectives are described next.

Sophistication of Strategy

The use of a particular model to solve a problem is evidence of the level of sophistication of the strategy and can indicate how students understand multiplicative situations. Read the problem in Figure 3.1. This problem is engineered to specifically elicit a student-constructed model for multiplication of one-digit numbers. In general, problems that are designed in this way elicit students' default model: the model they are most comfortable using when not influenced by context. The student responses in Figure 3.1 show the use of a range of visual models along the progression from least to most sophisticated. Solution A is an equal groups model. Solution B is an array. Solution C is an area model. Solution D is an open area model.

Figure 3.1 Students often rely on their default model when responding to a question that directly asks for a model.

Look at this equation

6 × 4 = 24

Draw a model that represents this equation.

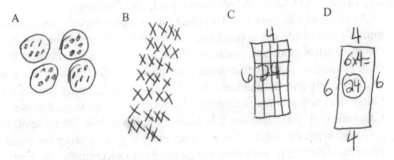

Instructional Guidance

The graphic in Figure 3.2 depicts the instructional path along the progression moving from equal groups to a mental model. Importantly, "visual models should be used as a way to understand and generalize mathematical ideas; that is, visual models are a means to the mathematics, not the end" (Petit et al., 2015, p. 3).

Figure 3.2 Hypothesized instructional path when using visual models to build multiplicative fluency and understanding (Hulbert, Petit, & Laird, 2015).

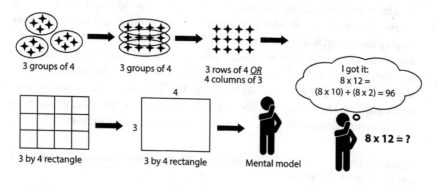

Equal Groups

Most students begin modeling multiplication by drawing equal groups models (Figure 3.1A). This is typical because it is a model of multiplication that is most closely related to repeated addition, which is a strategy most students first use to solve multiplication problems. Even the vocabulary we use when describing this model sounds like addition (e.g., 4 objects in 3 groups). Although this is a helpful beginning model for students to understand the importance of equal groups, it is limiting in its use and generalizability when factors increase in magnitude and as students expand and deepen their understanding of the properties of operations for multiplication. Teachers, therefore, need to help students move to a more useful model.

Using Arrays to Transition from Equal Groups

"A two-dimensional array is a rectangular arrangement of things into (horizontal) rows and (vertical) columns, such that each row has the same number of things and each column has the same number of things" (Beckmann, 2014, p. 141). In an array each dimension is represented by a number of discrete objects (objects that are not connected to each other). At first students may focus on the sets of objects in each row or column, but the array can also be used to help transition students to focus on the dimensions rather than the equal groups. This is an important instructional step for many students because it directly connects the more familiar equal groups model to one that expands the meaning of multiplication. Although equal groups can be seen in the rows and columns of an array (see Figure 3.3), the goal is to move students from using the equal groups to using the dimensions in the array as the factors in a multiplication problem (see Figure 3.4). This rectangular arrangement in an array and the move to connect the dimensions of the array to factors is important because it begins to set the stage for using area models. Area models are vitally important to students' understanding of the properties of operations that underpin fluency with multiplication.

The transition from equal group models to array models can include a point where students only consider one dimension of the array and as a result may display their objects in an array but still see the equal groups in the array only along one dimension. For example, Lainey's work in Figure 3.3 provides evidence that she is in the process of making the transition from an equal group model to an array model. She organized groups in an array; however, the evidence suggests she is only considering one dimension of her array or she may not be seeing dimensions at all, but rather groups organized in rows.

Figure 3.3 Lainey's solution. The evidence suggests that Lainey is considering the equal groups along one dimension.

Look at this equation

6 × 4 = 24

Draw a model that represents this equation.

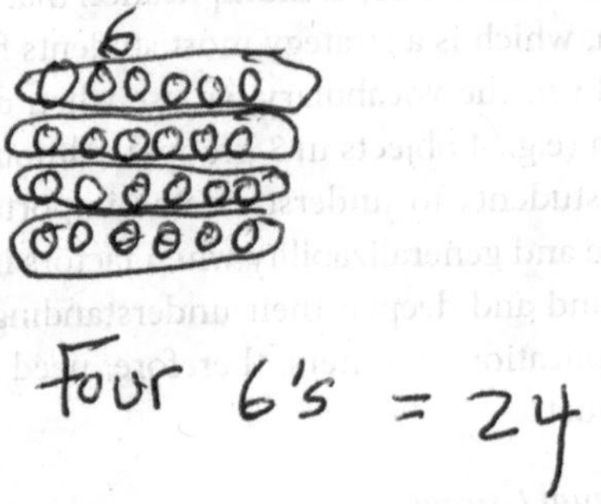

In contrast, in Figure 3.4 Kelyn is considering both dimensions of the array, not simply the equal groups.

Figure 3.4 Kelyn's response. Kelyn's solution is focused on the dimensions.

How many wheels are there in 4 tricycles?

It is noteworthy here that the transition from seeing equal groups in an array along one dimension to seeing both dimensions of the array can present some challenges for students. In particular, students often have difficulty realizing that the dimensions and not the items themselves represent the factors of a multiplication problem.

In general, as students interact with arrays, they move through four stages as shown in Figure 3.5 (Battista, Clements, Arnoff, Battista, & Borrow, 1998). These stages are not developmental stages that teachers need to progress students through. Rather, they include errors (Stages 1 and 3) or developing understandings (Stage 2) evidenced in the use of arrays until students visualize and use both dimensions (Stage 4). Each of the stages is described in Figure 3.5.

Figure 3.5 Stages in development of use of arrays.

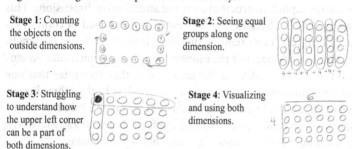

Stage 1: Counting the objects on the outside dimensions.

Stage 2: Seeing equal groups along one dimension.

Stage 3: Struggling to understand how the upper left corner can be a part of both dimensions.

Stage 4: Visualizing and using both dimensions.

Note that in Stage 3, students struggle to understand how the upper-left corner of an array can be a part of both the column and the row of the array. This is evidenced in a couple of ways. Students might express confusion about the object in the upper-left corner and why that piece appears to be double counted and/or they may neglect an entire row or column in the array trying to avoid what they think is a double count. In the example shown in Figure 3.5 Stage 3, a student might see that there are 6 circles in each row but think there are only 3 rows because they have already counted the corner piece.

Area Models Strengthen Focus on Factors as Dimensions

The next step in the progression of visual models for multiplication is for students to build area models from an understanding of arrays. In an area model, the factors represent continuous lengths rather than discrete amounts. There are many similarities between an area model and an array model. One similarity is that the number of "tiles" in the area model can be counted to find the product, just like the number of items can be counted in an array model. This feature helps some students more easily transition from the array to the area model. An array can be composed of any object (e.g., apples, circles, marbles), whereas area models are composed of continuous square units as shown in Figure 3.6. This difference plays a key role in helping focus attention on dimensions as factors.

Figure 3.6 An array and area model representing 3 × 4. In a discrete model the items are individual unconnected objects, as illustrated in the array on the left. In a continuous model the items are connected.

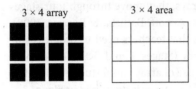

To help students transition from seeing rows and columns in an array or area model requires explicit instruction focusing attention on dimensions. This is easier in an area model because the area model is composed of continuous units, not discrete objects. The attention, therefore, is on determining the length of the dimensions of each side, not the number of squares on each side. As students develop this understanding of the area model, they recognize that one factor refers to one dimension and the other factor refers to the other dimension as shown in Figure 3.7. In this example the dimensions are 3 units × 4 units and the product is 12 square units. Note that the focus is drawn away from the corner tile to the lengths of the edges of the tiles of each dimension.

Figure 3.7 Area model 3 units × 4 units with an area of 12 square units.

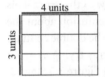

The Open Area Model Is the Final Bridge to Fluency

The open area model is the next step in the progression. It is the most flexible of the models and allows for work with larger numbers and ultimately with fractions, decimals, and algebraic expressions. Use of the open area model bridges place value understanding with the distributive property to support more efficient multiplicative strategies.

For example, in Figure 3.8 there is evidence that Haley knew the number 16 is equal to 10 + 6. This knowledge allowed Haley to use the distributive property to determine the product of 9 ×16 by distributing the 9 across (10 + 6) in the area model [(9 × 10) + (9 × 6) = 144].

The area model can help students develop strategies for multiplication that are based on understanding of place value as exemplified in Haley's solution when she decomposed 16 into 10 and 6. Notice also that the proportionality of the partitioned area model provides a visual clue as to the relative magnitude of each of the products. Ultimately, the goal is for students to move away from reliance on the open area model to flexible use of efficient algorithms that are based on place value and the distributive property.

Figure 3.8 Haley's solution. Haley used the open area model to solve the multiplication problem.

Sarah and Beth are saving beads to make bracelets.

Sarah has 9 beads.

Beth has 16 times as many beads as Sarah.

How many beads do Beth and Sarah have altogether?

Show your work.

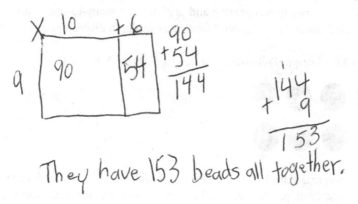

They have 153 beads all together.

Go To See Chapter 4: The Role of Concepts and Properties, Chapter 8: Understanding Algorithms, and Chapter 9: Developing Math Fact Fluency for a more in-depth discussion on how the open area model can be used to develop place value understanding, support understanding of the distributive property, and develop understanding and fluency of both math facts and algorithms.

The Role of Subitizing and Quick Images

Teachers often ask how to effectively use visual models to help students develop understanding of, and fluency with, multiplication. For example, how can I help students transition from using an equal groups model to an array or area model? Of course, there is not one right answer to this question. However, researchers indicate that one way to help is by building and capitalizing on subitizing and the use of quick images (Clements, 1999).

Subitizing is the ability to quickly identify the number of items in a small set without counting each object. There are two types of subitizing: perceptual and conceptual. Perceptual subitizing involves "recognizing a number without using any other mathematical process" (Clements, 1999, p. 401). For example, very young children can recognize an organized group of two to four objects without counting. Conceptual subitizing, unlike perceptual subitizing, involves mathematical processes. For example, a student may recognize that the number of dots in the pattern in Figure 3.9 consists of 8 dots by seeing that each row in the pattern has 4 dots and that there are two groups of four or 4 + 4 or 2 × 4. The ability to see each row as having 4 dots without counting the dots in each row is evidence of unitizing; that is, seeing one row as 4 dots and two rows as 2 rows of 4 dots. This ability to unitize is critical in the development of multiplicative reasoning and can be supported through the development of conceptual subitizing. Specifically, unitizing is critical in helping students move beyond counting by ones to recognizing and applying the many-to-one relationships in multiplication. (See Chapter 4 for a more in-depth discussion on unitizing.)

Figure 3.9 Conceptual subitizing—the student might see 2 rows of 4 circles.

"Conceptual subitizing must be learned and, therefore fostered, or taught" (Clements, 1999, p. 402) and can be used to help students transition to more sophisticated strategies and mental models. The use of quick images is one instructional strategy to build conceptual subitizing.

A quick image is best understood by providing an example. Imagine that you have the pattern in Figure 3.9 on a large piece of paper. You explain to the students that you are going to hold up the pattern for only a few seconds. Their job is to determine the number of dots in the pattern without counting each one. This promotes the use of mental images so important to developing more sophisticated strategies. Students may see two rows of 4 dots, or four columns of 2 dots, or two groups of 4 in the way that 4 is represented on dice. Examples of ways to use quick images to transition students from one visual model for multiplication to the next are described later.

Researchers indicate that subitizing is related to visualization (Markovits & Hershkowitz, 1997). Using quick images to develop conceptual subitizing combines visualization with subitizing as students describe their mental image of the pattern. Figure 3.10 contains a progression that can help you think about the type of quick images you can use with your students who are using counting-by-ones strategies.

Figure 3.10 Sample quick images going from familiar patterns with small numbers to equal groups to area models.

Source: (Clements & Sarama, 2014).

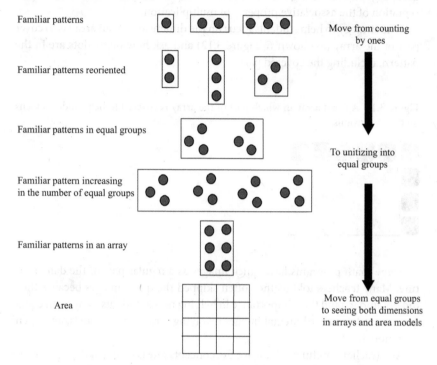

Using quick images is not a one-time event, but should be repeated over multiple times while increasing the number of objects in the familiar patterns until students can easily use their mental images to determine the number of objects in patterns like those in Figure 3.11.

Describing mental images with mathematical expressions can help students build understanding of various properties of operations. For example, a student might describe the number of dots in the 4 × 6 array in Figure 3.11 as 2 groups

Figure 3.11 These two different representations of the relationships in a 4 × 6 array help to illustrate the associative property.

$2 \times (4 \times 3) = 2 \times 12 = 24$ $(2 \times 4) \times 3 = 8 \times 3 = 24$

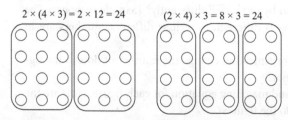

of 12 or $2 \times (4 \times 3) = 24$, whereas another student might describe the number of dots in the patterns as 3 groups of 8 or $(2 \times 4) \times 3 = 24$. Highlighting these groupings and recording the strategies with equations provides a visual representation of the associative property of multiplication.

One strategy to help students focus on the dimensions of an array is to cover part of the array (as shown in Figure 3.12) and ask how many dots are in the pattern, including the covered part.

Figure 3.12 A 6×4 array in which part of the array is covered to help students focus on the dimensions.

Some math programs have quick images as a regular part of the daily routine. Many teachers told us they often skipped the quick images because they did not understand their importance in helping move students away from equal groups models toward mental images involving arrays, area models, and open area models.

As a teacher, careful planning for the concepts one is trying to develop maximizes the impact of using quick images. For example, if a teacher's goal is to develop understanding of the associative property, the teacher might use a 4×6 array as shown in Figure 3.11. As students generate different equations to represent the number of dots in the array, the teacher writes the equations on the board and sketches the images as shown in Figure 3.13. The teacher then has students study the equations and make observations using questions that help focus their attention on understanding the associative property.

Sample questions to focus attention on the associative property include the following:

1. What do you notice about the factors in each of these equations? [Sample response: They are the same numbers, but organized differently.]
2. What do you mean by organized differently? [Sample response: The order of the factors is different, as well as different numbers are in parentheses.]
3. Did this different organization change the number of dots in the figure? Why or why not?
4. Study the figures and matching equation for each figure. How are the equations related to the figures?

Because students have developed these equations from a visual model and studied the relationship between the equations and the images, they can build understanding that the associative property is not a trick or a definition to memorize, but rather a mathematical tool they can use when solving problems.

Figure 3.13 Equations and visual models on a white board representing some relationships in a 6 × 4 array. Displaying on the white board as shown provides the opportunity for students to study relationships between the equations and the visual models representing the equations.

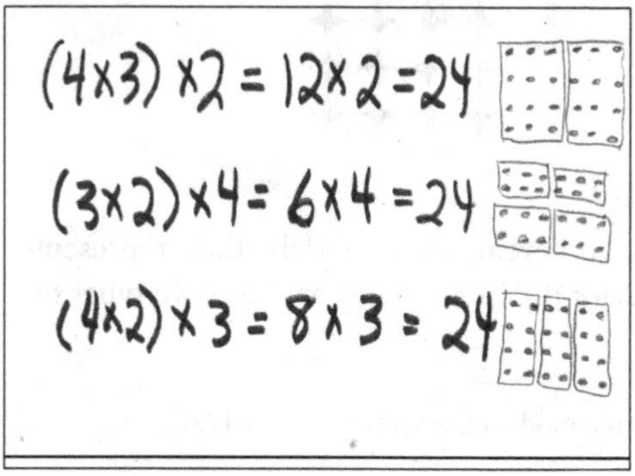

$$(4 \times 3) \times 2 = 12 \times 2 = 24$$

$$(3 \times 2) \times 4 = 6 \times 4 = 24$$

$$(4 \times 2) \times 3 = 8 \times 3 = 24$$

As mentioned in the introduction to this chapter, there are two important perspectives to consider when using the progressions in relation to visual models: 1) understanding evidence of the sophistication of student strategies when they solve problems and 2) guiding instructional decisions. The two perspectives work together. If a teacher knows the model a student is using, then they can use the progression as a guide to move the student's strategy and understanding to a more sophisticated level. We will continue to explore the role of visual models in the upcoming chapters.

Chapter Summary

- Visual models play a key role in developing multiplicative reasoning and fluency.
- Use of quick images combines subitizing and visualization. Together they help students transition to more sophisticated visual models and to mental models.

Looking Back

1. **Analyzing Common Visual Models for Multiplication:** In this chapter you became familiar with different visual models for multiplication. You learned that visual models for multiplication differ in appearance, sophistication, flexibility, and usefulness beyond whole number multiplication.

The following questions provide an opportunity for you to revisit ideas about visual models that were introduced in this chapter.

(a) In what ways are an equal groups model and an array alike? How are they different?

Figure 3.14 Equal groups model and array model representing 3 × 4 = 12.

Groups of Model   Array Model

(b) Figure 3.15 shows two different area models that represent 3 × 4 = 12. What is the major difference between these two types of models?

Figure 3.15 Area model and open area model representing 3 × 4 = 12.

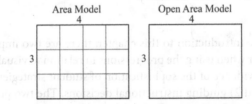

(c) Notice that the open area model is considered a *Transitional Strategy* on the *OGAP Multiplication Progression*, whereas the area model is considered an *Early Transitional Strategy*. Why is the open area model considered a more sophisticated visual model than the area model?

(d) Use Haley's strategy shown in Figure 3.8 to solve 16 × 19. How is an open area model that represents a 2 digit × 2 digit problem different from an open area model that represents a 1 digit × 2 digit problem? What would an open array model for a 3 digit × 3 digit problem look like?

2. **Conceptual Subitizing**: Conceptual subitizing involves using mathematical processes to determine a quantity. This differs from perceptual subitizing, which is the ability to recognize a number without using any mathematical process. For each of the following images, identify a specific mathematical process a student might use to determine the number of items in the image (Figure 3.16).

Figure 3.16 Conceptual subitizing images.

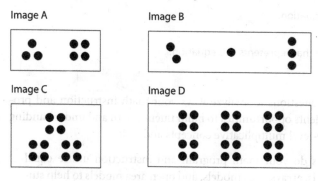

3. **Helping Students Transition to More Sophisticated Visual Models:**
Germain is comfortable using the array model for solving multiplication
problems like the one shown in Figure 3.17. The class is beginning to
operate with larger numbers, and his teacher, Ms. James, wants to help
him transition to a more efficient model and eventually to a multiplicative strategy for multiplication.

Figure 3.17 Sample of Germain's array model.

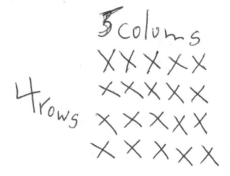

(a) What model might Ms. James use to help Germain transition to
next? Why did you choose this model?
(b) List some instructional strategies Ms. James can use to help Germain transition to the model you chose.

4. **Try This With Your Class:** Administer a question like the one in
Figure 3.18 to help you understand the types of visual models your students tend to use when solving multiplication problems. Sort the student
solutions into equal groups, arrays, area models, and open area models.
What did you find out about your students' use of visual models? What
instructional modifications will you make based on this information?

Figure 3.18 Sample question to gather evidence on students' use of visual models.

Look at this equation.

6 × 4 = 24

Draw a model that represents this equation.

Instructional Link

Use the following questions to analyze ways your math instruction and program provides students opportunities to build fluency with and understanding of important grade-level multiplicative concepts and skills.

1. In what ways do your math program and instruction utilize equal groups models, arrays, area models, and open area models to help students develop multiplicative reasoning and fluency?
2. How can you use quick images in your instruction to develop unitizing and understanding of the properties of operation?

The Role of Concepts and Properties

4

Big Ideas

- Linking the learning of arithmetic to important foundational mathematical ideas is fundamental to developing multiplicative reasoning and procedural fluency with understanding.
- Foundational mathematical ideas for developing flexibility and fluency when multiplying and dividing whole numbers include unitizing, place value understanding, and properties of operations.
- In addition to supporting the development of unitizing and place value, visual models can be used to highlight and deepen understanding of the commutative, associative, and distributive properties.

Continuing the focus on developing procedural fluency with understanding that began in the last chapter, this chapter focuses on three mathematical ideas that are foundational for developing flexibility and procedural fluency when multiplying and dividing whole numbers:

1. Unitizing
2. Place value understanding
3. Properties of operations

The meaning of procedural fluency was described in detail in Chapter 1. Because of its importance to the topics presented in this chapter, the essential components of procedural fluency are reiterated here.

Procedural fluency includes the following:

1. Knowledge of the steps in a procedure
2. Knowledge of when to use the procedures appropriately
3. Skill in applying a procedure flexibly, accurately, and efficiently
4. Underlying conceptual understanding of mathematical ideas

In this chapter, we focus on how to build conceptual understanding of the important mathematical ideas that underlie multiplication and division in order to develop procedural fluency.

CCSSM

The CCSSM supports this definition of the development of procedural fluency, as can be seen by the fact that place value and properties of operations play a prominent role throughout the standards that are focused on the development of multiplicative reasoning (See Table 2.1 in Chapter 2 and Table 8.1 in Chapter 8). The CCSSM does not expect mastery of standard algorithms for multiplication and division until after students have developed strategies based on place value, properties of operations, and the inverse relationship between multiplication and division (CCSSO, 2010).

The *OGAP Multiplication and Division Progressions*

Unitizing, place value understanding, and properties of operations are key to transitioning students from less to more sophisticated and efficient ways to solve multiplication and division problems, as discussed in more detail in Chapter 2. These concepts and properties are listed on the arrow on the right side of the progressions to illustrate the important role in moving students to more sophisticated strategies. This chapter examines these mathematical ideas and includes discussion of instructional strategies that focus on these concepts and properties to develop students' multiplicative reasoning, flexibility, and fluency.

Unitizing

When students first solve multiplication and division problems, they often bring additive reasoning and strategies to their solutions, such as counting by ones or repeated addition. Unitizing is a fundamental concept that allows students to transition from additive strategies toward more sophisticated multiplicative reasoning and strategies.

Unitizing is the ability to see a group, call it one group or unit, but also know that it is worth another value. For example, one tricycle can be understood as 1 group of 3 wheels. This idea that one group can simultaneously be 3 ones and 1 three is a new idea for students when they start to reason multiplicatively; it is the big idea of unitizing and it is foundational for the development of the concepts of multiplication and division (Fosnot & Dolk, 2001; Steffe, 1988; Ulrich, 2015).

In the problem "How many wheels on 29 tricycles?" one of the factors represents the number of groups (29 tricycles) and the other factor represents how many in each group (3 wheels per tricycle or per group). The number of groups and the number in each group have different associated units that are dependent upon each other: the total number of wheels is dependent upon the number of wheels on each tricycle. At the heart of multiplication are these equal groups, or composite units.

Students who are counting each object within a group to solve problems involving multiplication or sharing out by ones when solving division problems have not begun to unitize or conceptualize composite units. That is, they have not begun to simultaneously count individual objects and groups. See Figures 4.1 and 4.2 for examples of solutions where individual objects are counted by ones rather than groups.

Figure 4.1 Hunter's response. Hunter drew and counted each wheel by ones to find the total of 20 wheels as shown by the tick marks in each wheel.

> There are 5 cars in a parking lot. Each car has 4 wheels. How many wheels are there in all? Show your work.

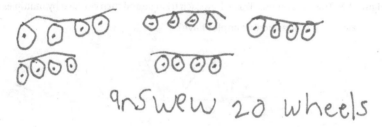

Figure 4.2 Diego's response. Diego's response has clear evidence that he distributed the cookies by ones.

> There are 12 cookies and 4 children. How many cookies will each child get if the cookies are shared equally among the 4 children?

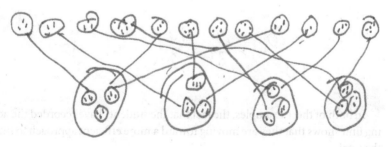

Focused instruction on unitizing can help students move beyond these *Early Additive* strategies of counting or sharing out by ones. Using quick images as discussed in Chapter 3 is one instructional strategy to help support students in making this transition to conceptualize groups. As students begin to unitize, they can think about each group as a quantity, allowing them to add that quantity repeatedly. However, their reasoning at this stage is still additive.

See Chapter 3 for more information about using quick images and their role in developing multiplicative reasoning and fluency.

As unitizing becomes internalized, students can move from repeated addition to skip counting or counting multiples of the number in each group, thereby keeping track of both the number in each group and the number of times it is repeated. The process allows for students to move from an *Additive* strategy to an *Early Transitional* strategy of skip counting. Often students initially draw or represent each object using a visual model to keep track of the number of multiples, as Tracy did in Figure 4.3. Students then transition from drawing each object to using a more efficient method to keep track of the groups, as reflected in Mark's solution strategy in Figure 4.4.

Figure 4.3 Tracy's response. Tracy drew each triangle and then counted by multiples of 3.

How many sides do 6 triangles have?

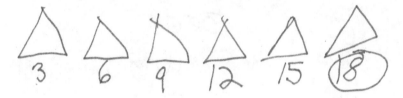

Figure 4.4 Mark's response. Mark's response shows evidence of skip counting by multiples of 3 without the use of drawing to keep track of the number of multiples.

How many wheels do 15 tricycles have?

3   6 9   12   15   18   21   24 27
30 33   36   39   42   45

In both of these examples, the fact that the students have recorded the accruing unit shows that they are moving toward a more efficient approach to finding the total.

Fluency with multiples is key to the transition from *Additive* to *Early Transitional* strategies. This involves both knowing the skip count sequence and understanding the concept of multiples, that 3 × 5, for example, means 3 multiples or groups of 5, or 15. Study solutions A and B in Figure 4.5. The evidence in solution A suggests a transition from skip counting to an organized list of the multiples of the heights of the boxes. In contrast, in Solution B the multiples of 12 and 14 are interpreted as a multiplication expression.

Figure 4.5 Two correct solutions to a problem designed to elicit strategies for solving problems involving multiples.

Abdi was stacking boxes that are 14 inches tall next to boxes that are 12 inches tall. Abdi wants to make both stacks the same height. At what height will the stacks be the same?

Solution A   Solution B

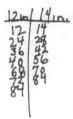

The 2 heights will be the same at 84 in or 6 boxes for the 14 in ones and 7 boxes for the 12 in boxes

$$\begin{array}{r} 12 \\ \times\ 7 \\ \hline 84 \end{array}$$

$$\begin{array}{r} 14 \\ 6 \\ \hline 84 \end{array}$$

By looking at both solutions side by side, students can make connections between the organized list of multiples to multiplicative thinking: there are 7 multiples of 12 or (12 × 7) and 6 multiples of 14 or (14 × 6).

Understanding the base-10 number system also requires understanding of unitizing. Fosnot and Dolk (2001) describe the prominent role of unitizing in the development of place value understanding.

Unitizing underlies the understanding of place value; ten objects becomes one ten. Unitizing requires that children use numbers to count not only objects but also groups—and to count them both simultaneously. The whole is thus seen as a group of a number of objects. The parts together become the new whole, and the parts (the objects in the group) and the whole (the group) can be considered simultaneously. For learners, this is a shift in perspective. Children have just learned to count ten objects, one by one. Unitizing these ten things as one thing—one group—requires almost negating their original idea of number (p. 11).

The base-10 number system is remarkably efficient for representing and calculating with numbers, but it is based on a multiplicative understanding that the position of a digit determines its value and the value of that position is always ten times the value of the place to its right.

At the most basic level, to understand the base-10 number system, students need to understand that 1 ten can simultaneously be 10 ones, and 100 can simultaneously be 10 tens and 100 ones, and so on. This understanding is extended as students decompose numbers. For example, in the number 73 the 7 does not mean 7 ones. Rather it means 7 groups of tens or 7 units of 10. Using base-10 representations is one way to make a direct link between unitizing and place value. Understanding the unitized nature of our number systems allows students to grasp the multiplicative relationships between the ones, tens, and hundreds blocks and, ultimately, to operate with larger numbers. Figure 4.6 illustrates how base-10 blocks are used to represent a three-digit number and help students move from unitizing to skip counting to more formalized place value understanding.

Figure 4.6 Base-10 blocks representing 262 that can be used to link skip counting and unitizing to place value.

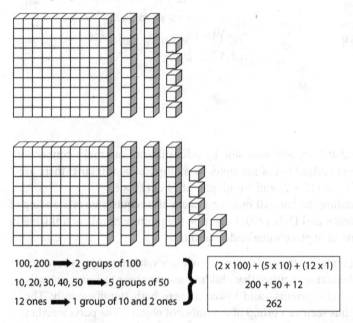

100, 200 ➡ 2 groups of 100

10, 20, 30, 40, 50 ➡ 5 groups of 50

12 ones ➡ 1 group of 10 and 2 ones

}

$(2 \times 100) + (5 \times 10) + (12 \times 1)$
200 + 50 + 12
262

The Role of Place Value

There are two interrelated aspects of student understanding of place value that help develop fluency and flexibility when multiplying and dividing whole numbers: 1) understanding place value of numbers and 2) understanding place value in computation for multiplication and division (Battista, 2012). To begin to understand how these two ideas are interrelated, study Pat's response in Figure 4.7. How did Pat use understanding of place value of the given factors to solve the problem? How did Pat use her understanding of place value in her computation?

Pat's solution shows evidence of using place value understanding of individual numbers when she correctly decomposed each of the factors and of applying place value understanding in computation when multiplying the factors and when adding the sum of the individual products. Pat's solution is an

Figure 4.7 Pat's response. Pat's response shows evidence of understanding place value of the numbers 64 and 12 and using understanding of place value to compute the solution.

John bought 12 boxes of crayons. Each box contained 64 crayons. How many crayons were there altogether?

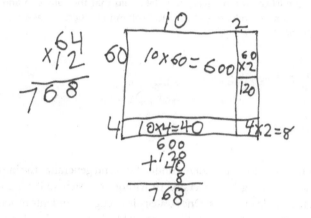

example of the open area model, which is more fully described and explained in Chapter 8.

There is a progression in the development of place value that begins with counting by ones, moving to unitizing by groups of ten, hundreds, thousands, etc., skip counting by place values to solve problems (e.g., $5 \times 10 = 50$ because 10, 20, 30, 40, 50); and then operating on numbers by decomposing and combining numbers (Battista, 2012). Decomposition by place value allows students to use *Transitional* strategies such as Pat's use of the open area model in Figure 4.7 and finally to solve multiplication and division of multidigit numbers using efficient algorithms at the *Multiplicative* level.

Multiplication by powers of 10 (e.g., $100 \times 17 = 1700$) is another important place value concept. This concept is often poorly understood and in some cases taught simply as a rule. That is, $100 \times 17 = 1700$ because there are two zeros in 100, and so to find the product you add two zeros to the other factor (17). However, adding zero does not change the value of a number (e.g., $17 + 0 = 17$), so this rule is mathematically incorrect. Sometimes students are told instead to annex or attach zeros at end of the factor for each power of ten. Although this is a mathematically correct rule for whole numbers, it expires in grade 5 when students start multiplying and dividing decimals by powers of 10 (e.g., $2.5 \times 100 \neq 2.500$) (Karp, Bush, & Dougherty, 2014).

One should not underestimate the impact of the misconception of adding zeros for each power of 10. In a study conducted by OGAP involving 47 fifth grade students, 62 percent of the responses to the following OGAP item showed evidence of this misconception as shown in Tom's response in Figure 4.8.

Figure 4.8 Tom's response. Tom's response shows evidence of the misconception of adding zeroes to find a product of a power of 10.

Jake was asked to explain the rule for the following function machine. He noticed that the numbers in the OUT column had one more zero than the number in the IN column on every row. Jake said that the rule is to add a 0 to the IN column numbers to get the OUT column numbers. Is Jake correct? Why or why not?

IN	OUT
1	10
10	100
100	1000
1000	10000
10000	100000

Jake is correct because 1 add 0 = 10
and the number in the OUT is going to be
the next IN.

The issue then, is how to help students understand and generalize the impact of multiplying and dividing by powers of 10 without overgeneralizing a rule that only applies to whole numbers. One strategy is to engage students in activities that involve linking visual models like base-10 blocks to verbal and written words, equations, and place value charts, as is shown in Figure 4.9. Doing this

Figure 4.9 In the problem 100 × 3 = 300 the 3 is shifted two places to the left for each power of 10.

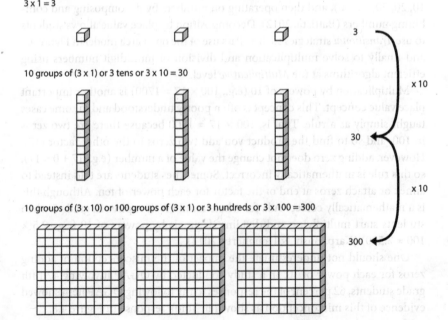

$3 \times 1 = 3$

10 groups of (3 x 1) or 3 tens or 3 x 10 = 30

×10

30

10 groups of (3 x 10) or 100 groups of (3 x 1) or 3 hundreds or 3 x 100 = 300

×10

300

with multiple examples allows students to see the pattern that the digits are actually shifting to the left for multiplication for each power of 10 and to the right when dividing by powers of 10.

As stated earlier, this notion of digits shifting when multiplying or dividing by powers of 10 is important because it holds true for multiplication of decimal fractions. For example, in the problem 8.1×100 the digits shift two places to the left, just like in the whole number example shown earlier.

$$8.1 \times 100 = 810$$

810 can also be thought of as 8.1 hundreds; the zero indicates that there are no groups of one. Student understanding of multiplying by multiples of powers of 10 should be logically built upon the notions described earlier. Some examples of numbers that are multiples of powers of 10 include 30, 400, and 8,000. For example, using place value understanding, 5×60 can be expressed as 5×6 tens or 30 tens or $(5 \times 6) \times 10$. In all three cases we can reason that 30 tens are equal to 300. One way to help students understand this idea is by having them work through a sequence of problems using base-10 representations and linking verbal and written language to the equations. Study the sequence of problems that follow. What patterns and relationships do you notice? In particular notice what is happening to the place value of the product as you multiply by powers of 10 and multiples of the power of 10.

Sample sequence:

1. $3 \times 40 = 3 \times 4 \times 10 = 12$ tens or 120
2. $3 \times 400 = 3 \times 4 \times 100 = 12$ hundreds or 1,200
3. $3 \times 4,000 = 3 \times 4 \times 1,000 = 12$ thousands or 12,000
4. $3 \times 40,000 = 3 \times 4 \times 10,000 = 12$ ten-thousands or 120,000
5. $80 \times 50 = (8 \times 10) \times (5 \times 10) = (8 \times 5) \times (10 \times 10) = 40$ hundreds $= 4,000$

Note that this last example involves both the commutative and associative properties, which are discussed in the next section. It may seem easier to simply provide students a series of rules to follow for multiplying by powers of 10 or by multiples of powers of 10. However, the conceptual understanding described earlier is vital for students to develop fluency and flexibility for reasoning and solving a variety of multiplication and division problems. In addition, knowledge based on conceptual understanding is longer lasting than rule-based notions (Hiebert & Carpenter, 1992).

In contrast to Tom's response in Figure 4.8, Mia's response in Figure 4.10 shows evidence that she understands the multiplicative relationship between the values in the IN column and the values in the OUT column. In addition, she seems to realize that adding zero to a number does not change the magnitude of the number. Her response shows understanding of the additive identity property of zero.

Figure 4.10 Mia's response. Mia's response shows evidence of understanding the multiplicative relationship between the IN and OUT in the function machine.

Jake was asked to explain the rule for the following function machine. He noticed that the numbers in the OUT column had one more zero than the number in the IN column on every row. Jake said that the rule is to add a 0 to the IN column numbers to get the OUT column numbers. Is Jake correct? Why or why not?

IN	OUT
1	10
10	100
100	1000
1000	10000
10000	100000

Jake is incorrect because 1+0=1 not 10 the real rule is 10×In colum.

Mia's correct response, as well as the sample problem sequence earlier, leads us to the focus of the next section: using properties of operations to help build procedural fluency with flexibility and understanding.

Properties of Operations

The commutative and associative properties of addition, together with the other properties of arithmetic (which include the commutative and associative properties of multiplication and the distributive property) form the building blocks of all of arithmetic. Ultimately, every calculation strategy, whether a mental method of calculation or a standard algorithm, relies on these properties. These properties allow us to take numbers apart, to break arithmetic problems into pieces that are easier to solve, and to put the pieces back together. The strategy of decomposing into simpler pieces, analyzing the pieces, and then putting them back together is important at every level of mathematics and in all branches of mathematics (Beckmann, 2014, p. 100).

Many students can use their intuitive understanding of the properties of operations to develop flexible strategies and structure their thinking when multiplying and dividing whole numbers. However, because it is known that not all students engage with the properties intuitively (Empson & Levi, 2011), it is important that teachers provide students rich and focused instruction that engages students in both understanding the properties and using this understanding to flexibly solve multiplication and division problems. Table 4.1 provides a list of the properties and relationships relevant to multiplication and division that will be discussed throughout this section. It is not important that students possess a formal definition of the properties. Much more important is that students understand the properties and know how to use them to make sense of multiplication and division problems and to flexibly solve problems with a variety of contexts and numbers. Table 4.1 is followed by examples of ways to engage students in reasoning through instructional strategies such as questioning to highlight properties when students share solutions, using engineered formative assessment questions, and making and testing conjectures.

Table 4.1 Properties and relationships – multiplication and division.

Properties		
Multiplicative identity property	$a \times 1 = a$ The product of any number and 1 is that number.	$35 \times 1 = 35$ $35 = 35$
Commutative property of multiplication	$a \times b = b \times a$ When two numbers are multiplied together, the product is the same regardless of the order of the factors.	$25 \times 2 = 2 \times 25$ $50 = 50$
Associative property of multiplication	$(a \times b) \times c = a \times (b \times c)$ When three or more numbers are multiplied, the product is the same regardless of the grouping of the factors.	$(3 \times 5) \times 4 = 3 \times (5 \times 4)$ $15 \times 4 = 3 \times 20$ $60 = 60$
Distributive property of multiplication over addition	$a \times (b + c) = (a \times b) + (a \times c)$ The sum of two numbers times a third number is equal to the sum of the products of each of the addends and the third number.	$4 \times (6 + 2) = (4 \times 6) + (4 \times 2)$ $4 \times 8 = 24 + 8$ $32 = 32$
Relationships		
Inverse relationship between multiplication and division	If $a \times b = c$, then $c \div b = a$, and $c \div a = b$ There is an inverse relationship between multiplication and division.	$4 \times 6 = 24$ $24 \div 6 = 4$ $24 \div 4 = 6$
Multiplication by zero	$a \times 0 = 0$ Any number times zero is zero.	$5 \times 0 = 0 = 0 \times 5$

Highlighting Properties in Student Solutions

As noted earlier, the CCSSM states that students in grades 3–5 should solve multiplication and division problems with strategies that are based on place value, properties of operations, and/or the relationship between multiplication and division.

Study Olivia's response in Figure 4.11. Notice that Olivia used her understanding of place value and the distributive property of multiplication over addition. Although we don't know if this third grade student had any formal instruction in the use of the distributive property, the evidence suggests that

Olivia decomposed 19 and then used her understanding of multiplication to multiply 15 ft. × 10 ft. and 15 ft. × 9 ft. to solve the problem.

Figure 4.11 Olivia's response. There is evidence in the response of decomposing 19 and then applying the distributive property.

How many 1-foot square tiles does it take to cover the school playground?

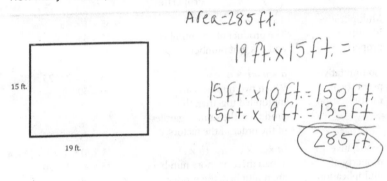

Notice that Olivia used units throughout her solution. However, the final unit in 285 ft. is incorrect and should be square feet (ft^2), which could be easily illustrated with the given area model. The importance of the meaning of the quantities in multiplication and division problems is addressed in Chapter 5.

Whereas some children intuitively use the distributive property, others need more explicit instruction. The explicit instruction is not about teaching the property directly and having students practice that strategy, but rather providing platforms upon which to have discussions that engage students in understanding how to use the underlying concepts in the property to flexibly multiply and divide numbers. For example, a teacher might capitalize on Olivia's solution by sharing it with the entire class and asking probing questions that deepen all students' understanding of the distributive property. Her solution can be further illustrated on the model provided in the problem.

Sample probing questions about Olivia's solution:

1. Is Olivia's solution correct or incorrect? Why or why not?
2. How did Olivia solve the problem?
3. Can Olivia's strategy be used to solve other multiplication problems like 5 × 34? Why or why not?
4. Name some other multiplication problems that can be solved using this strategy.
5. Use the visual model to illustrate Olivia's strategy.

In the example shown in Figure 4.12, Omar found the product of 25 × 6 by first finding the product of 24 × 6. Although we don't know why he chose to do this, it seems that he knew that the product of 24 × 6 was equivalent to the product of 12 × 12 through a strategy of doubling and halving the factors.

Figure 4.12 Omar's response. Omar's solution involves a doubling and halving strategy, which illustrates the distributive, commutative, and associative properties.

Samantha's class has 25 bags of cookies. Each bag contains 6 cookies. How many cookies does Samantha's class have all together?

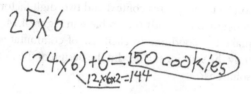

Notice that to find the product of 25 × 6, Omar used the distributive property to change the product into (24 × 6) + 6. To find 24 × 6, he first halved the 24 and then doubled the 6 to get 12 × (6 × 2) or 12 × 12. By breaking down his solution into steps, we can see the distributive, commutative, and associative properties in this solution strategy:

$$25 \times 6 = (24 \times 6) + 6 \quad \text{Distributive property}$$
$$= (12 \times 2) \times 6) + 6$$
$$= (12 \times (6 \times 2) + 6 \quad \text{Associative and commutative properties}$$
$$= (12 \times 12) + 6$$
$$= 144 + 6$$
$$= 150$$

In addition to asking probing questions about Omar's strategies such as those shown earlier for Olivia's solution in Figure 4.11, teachers can illustrate the properties visually with the open area model. In Figure 4.13, the associative property is illustrated by an open area model for 6 × 24. The shaded area has been moved, but the total area, or the product, remains the same. This model

Figure 4.13 The doubling and halving strategy and the associative property can be illustrated with open area models.

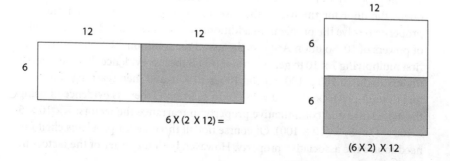

can help students reason that a doubling and halving strategy like Omar used can work for all numbers because of the associative property.

Tasks Engineered to Highlight Properties of Operations

Teachers may also design or select questions that provide the opportunity to use the different properties. For example, the problem in Figure 4.14 was engineered to elicit use of the commutative property and/or the associative properties of multiplication by using an area context and two-digit factors that can be decomposed into tens and ones. Study the problem in Figure 4.14. What aspects of the problem make it a good one to elicit use of commutative and associative properties? Study the solutions. How did the design of the problem allow for these different solutions?

Figure 4.14 Correct responses to the problem using different strategies and different levels of place value understanding.

Mr. Jones ordered 7 cases of paper. There are 10 packages of paper per case. Each package contains 500 sheets of paper. How many sheets of paper did he order?

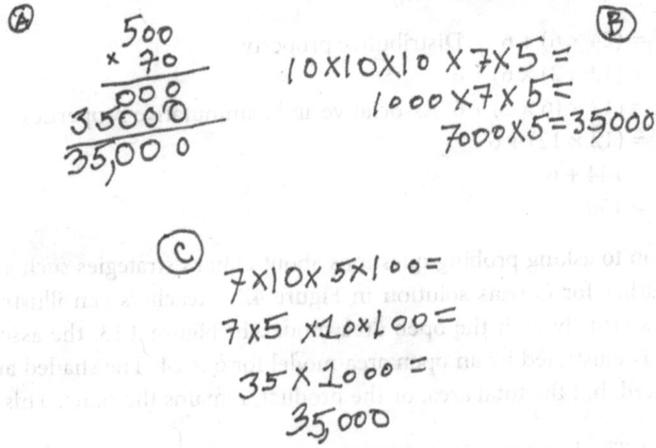

Ⓐ
$$500 \times 70$$
$$35000$$
$$35,000$$

Ⓑ
$$10 \times 10 \times 10 \times 7 \times 5 =$$
$$1000 \times 7 \times 5 =$$
$$7000 \times 5 = 35000$$

Ⓒ
$$7 \times 10 \times 5 \times 100 =$$
$$7 \times 5 \times 10 \times 100 =$$
$$35 \times 1000 =$$
$$35000$$

You probably noticed that the problem has three factors. This fact alone opens the door for the use of the associative property and the commutative property to solve the problem. In addition, the factors include 10 and multiples of powers of 10. Solution A shows the use of the traditional US algorithm after first multiplying 7×10 to get 70. In solution B there is evidence that the student understood that 500 is 100×5 and 100 is 10×10 and then rearranged the factors to multiply $10 \times 10 \times 10 = 1,000$. In solution C there is evidence of using the associative and commutative property to rearrange the factors: $7 \times 10 \times 5 \times 100 = (7 \times 5) \times (10 \times 100)$. Of course not all three-factor problems elicit the need to use the associative property. However, both the order of the factors in the problem and the fact that the problem involves numbers that are multiples of powers of 10 provides opportunities to draw on understanding of these properties to enact a flexible strategy to find the product of the three numbers.

To help students understand the underlying properties in these solutions, teachers can share the solutions with the class and have students explore the relationships between the solutions using probing questions.

Sample questions to pose about the student solutions in Figure 4.14 include the following:

1. How did each student solve this problem?
2. How are the solutions related to each other?
3. Do the calculations in the solution need to be completed in the same order that the numbers are presented in the problem? Why or why not?
4. Where do you see the 500 in solutions B and C?
5. Where do you see the 70 in solutions B and C? Is this okay to do? Why or why not?
6. Write another three-factor multiplication problem and solve it using the strategy demonstrated in solutions B and C. Explain why using this strategy makes the calculations easier to carry out.

It is also important to give students tasks that specifically target understanding of a certain concept or property. Study the problem and solutions in Figure 4.15. How was the problem designed to elicit understanding of the commutative property of multiplication or multiplication by zero? What do you notice about students' understanding of these properties in solutions A, B, and C?

Figure 4.15 Problem designed to elicit application of the commutative property of multiplication or the impact of multiplication by zero.

Complete the following equation to make the statement true.

30 x _____ = 71 x _____

A

1) 30 x **71** = 71 x **30**

B

1) 30 x **71** = 71 x **30**

$7 \times 10 = 710$
$30 \times 10 = 300$
$71 \times 30 = (2130)$
$30 \times 30 = 900$

$71 \times 20 = 1420$
$30 \times 20 = 600$
$71 \times 40 = 2840$
$30 \times 40 = 700$
$30 \times 50 = 1500$
$30 \times 60 = 1800$

$71 \times 50 = 2210$
$21 \times 60 = 2920$

$30 \times 71 = (280)$

C

1) 30 x **0** = 71 x **0**

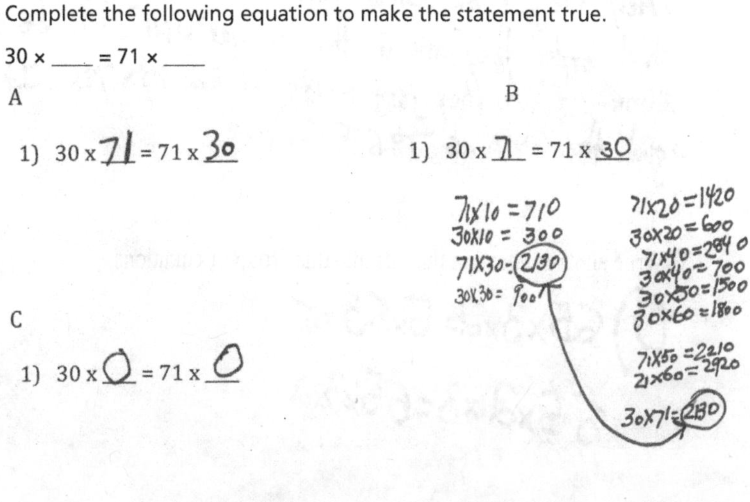

The open number sentence in the question in Figure 4.15 provided the opportunity for students to complete the equations using their understanding of either the commutative property or multiplication by zero. Solutions A and C have correct responses based on either multiplication by zero or the commutative property of multiplication. Solution B, however, shows evidence of having to carry out multiple computations to ultimately determine that $30 \times 71 = 71 \times 30$.

Another strategy for highlighting properties of multiplication is to ask students to make and test conjectures by determining whether equations are true or false (Carpenter et al., 2003). In the task shown in Figure 4.16, students are given equations and asked to explain why they are true. In Solution A, the student recognizes the associative and commutative properties and is able to generate different true equations. In Solution B, the student had to perform the calculations to check, but then was able to generalize about why the products were equal. Ultimately, students should be able to recognize and use the properties to justify these equations without having to carry out calculations.

Figure 4.16 Two solutions to a problem engineered to engage students in making and testing conjectures.

Review the following true equations.

(a) $786 \times 5 \times 2 = 10 \times 786$

(b) $5 \times 20 \times 3 = 3 \times 5 \times 20$

(c) $45 \times 5 \times 4 = 4 \times 5 \times 45$

Solution A

1) Write a general statement as to why these equations are true.

They are just switching the numbers around & no matter what order they are in the answer will still be the same. Also A, they just multiplied 5x2=10x786. It still equals the same as 786x5x2.

2) Write another equation that fits into this group of equations.

0) $65 \times 3 \times 2 = 6 \times 65$ or

$65 \times 2 \times 3 = 65 \times 3 \times 2$

Figure 4.16 Continued.

Solution B

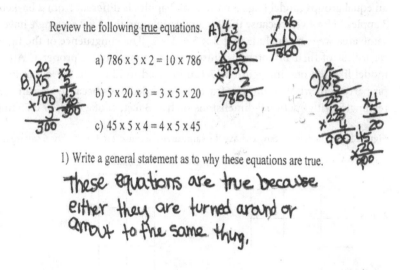

Review the following <u>true</u> equations.

a) 786 x 5 x 2 = 10 x 786

b) 5 x 20 x 3 = 3 x 5 x 20

c) 45 x 5 x 4 = 4 x 5 x 45

1) Write a general statement as to why these equations are true.

These equations are true because either they are turned around or amout to the same thing,

2) Write another equation that fits into this group of equations.

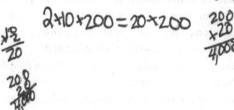

Representing Properties of Multiplication with Visual Models

Visual models are a powerful way to illustrate the commutative and associative properties of multiplication, the distributive property of multiplication over addition, and the inverse relationship between multiplication and division. This section provides examples of ways to use visual models to help your students deepen their understanding of these three properties.

Commutative Property of Multiplication

The commutative property of multiplication states that when two numbers are multiplied together, the product is the same regardless of the order in which the factors are multiplied. The generalized case of the commutative property can be expressed as $a \times b = b \times a$. The area models in Figure 4.17 illustrate this property because the areas of the two rectangles are congruent even though each rectangle is oriented differently. Note that it is harder to see this commutativity in an equal groups model (e.g., 3 baskets of 2 apples is different from 2 baskets of 3 apples.) However, because the 2 × 3 rectangle and the 3 × 2 rectangle have the same area we can conclude that 2 × 3 = 3 × 2. The congruence of the figures, regardless of their orientation, illustrates the commutative property. An area model like the one in Figure 4.17 can be used to illustrate the commutative property with whole numbers, fractions, and decimals because we can create rectangles with fractional dimensions such as 3 feet, ¾ of a yard, or 0.75 miles.

Figure 4.17 The two rectangles have the same areas because 2 × 3 = 3 × 2 or 6 square units.

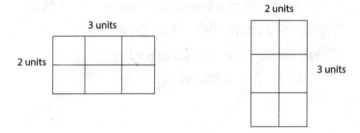

Many math programs use the convention that in a multiplication expression the first factor represents the number of rows and the second factor represents the number of columns. This is a convention, rather than a rule, because mathematically arrays and area models are congruent when rotated.

To extend this specific case to the generalized understanding of the commutative property of multiplication, one can assign the dimensions *a* units and *b* units to each rectangle; *a* and *b* can be any positive rational number and any linear dimension unit (e.g., inches, mm) (see Figure 4.18). In both rectangles, the value for *a* is the same and the value for *b* is the same. In this case, *a* represents the length of the shorter side of each rectangle and *b* represents the length of the longer side of each rectangle. The area, therefore, is the same in both rectangles. That is, the product of the factors is unaffected by the order in which they are multiplied ($a × b = b × a$).

Figure 4.18 If the lengths of *a* and *b* are positive numbers, then $a × b = b × a$ and the areas of each of the rectangles is *ab* square units.

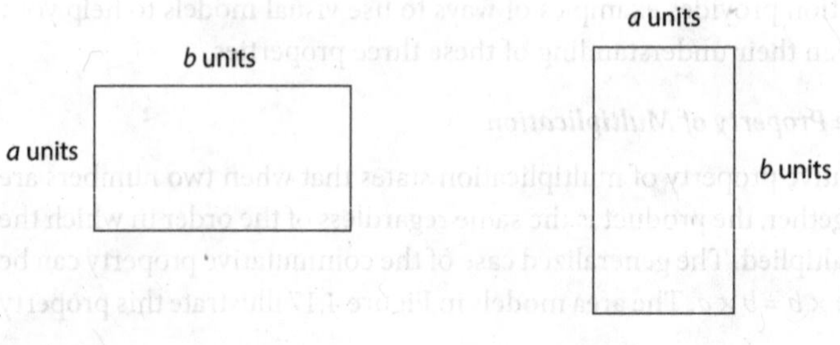

Using quick images of arrays or area models, such as those suggested in Chapter 3, is another effective way to highlight the commutative property. For the 3 by 5 array shown in Figure 4.19, for example, one student might see 3 groups of 5 and another 5 groups of 3. The teacher can represent this as $5 \times 3 = 3 \times 5$.

Figure 4.19 Quick images can be used to generate the commutative property through equal groups in an array.

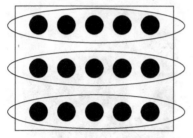
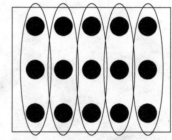

5 groups of 3 or (5 x 3)   3 groups of 5 or (3 x 5)

The Associative Property of Multiplication

This same line of thinking can be applied to understanding the associative property of multiplication. Study the rectangular prisms in Figure 4.20. Notice that all three prisms have a volume of 24 cubes regardless of their orientations. To help understand the associative property think of the total volume in terms of the number of 1 unit $\times$ 1 unit $\times$ 1 unit cubes that fill a given volume.

The prism can be oriented three different ways, each illustrating a different base.

- A is oriented so that the side with dimensions $2 \times 4 \times 1$ is the base
- B is oriented so that the side with dimensions $2 \times 3 \times 1$ is the base
- C is oriented so that the side with dimensions $3 \times 4 \times 1$ is the base

In each case the total volume of the prism can be thought of as the area of the base of the prism, or one layer of cubes, multiplied by the height of the prism. Thus:

- A shows $(2 \times 4) \times 3 = 24$ cubic units
- B shows $(2 \times 3) \times 4 = 24$ cubic units
- C shows $(3 \times 4) \times 2 = 24$ cubic units

In each case notice that the expression in parentheses is the area of the base of the prism and the other factor is the height of the prism, which can also be thought of as the number of layers or iterations of the base needed to make the prism. This example illustrates that the three factors can be associated in different ways while preserving the product: 24 cubic units. Note that the example also involves using the commutative property of multiplication as the order of the factors were changed.

Figure 4.20 Rectangular prisms illustrating the associative property of multiplication with an example.

(a) (2 × 4) × 3 = 8 × 3 = 24 cubes

(b) (2 × 3) × 4 = 6 × 4 = 24 cubes

(c) (3 × 4) × 2 = 12 × 2 = 24 cubes

Generalizing this idea, the two rectangular prisms in Figure 4.21 have the same dimensions (a, b, and c) and volume, but are oriented differently. Different orientations of the same prism illustrate that regardless of how the dimensions are associated or the order in which you multiply the dimensions, the volume remains the same.

Figure 4.21 Rectangular prisms illustrating the generalized representation of the associative property of multiplication.

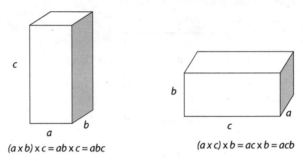

$(a \times b) \times c = ab \times c = abc$

$(a \times c) \times b = ac \times b = acb$

 Go To See Chapter 3: The Role of Visual Models for an example of how to use quick images to highlight the associative property of multiplication.

The Distributive Property of Multiplication

Initial development of understanding of the distributive property can be built using quick images (see Chapter 3) as students describe the strategies they use to determine the total number of objects in an array. For example, students might describe how they determined the total number of dots in the pattern in Figure 4.22 by saying, "I saw 4 groups of 2 and 4 groups of 3 and added them together." As students describe their solutions, writing multiplication equations to represent their descriptions can help students develop conceptual understanding of the distributive property.

Figure 4.22 The equation represents the student's description ("I saw 4 groups of 2 and 4 groups of 3 and added them together") using the distributive property.

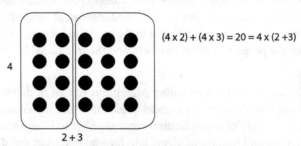

$(4 \times 2) + (4 \times 3) = 20 = 4 \times (2 + 3)$

Figure 4.23 illustrates the generalized representation of the distributive property. In this case $a + b$ is the decomposed length of the longest side of the rectangle, and c is the length of the shortest side. Distributing c over each of the additive parts $(a + b)$ results in the total area of the open area model: $(c \times a) + (c \times b) = ca + cb$.

Figure 4.23 An open area model illustrating the generalized representation of the distributive property of multiplication over addition.

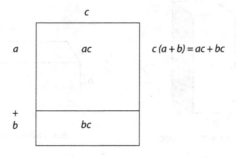

Further development of the distributive property of multiplication over addition is directly linked to the area and open area models used to transition students from *Additive Strategies* to *Multiplicative Strategies* with both whole numbers and fractions, and then again in middle and high school when multiplying algebraic terms. Initially, use of the area model helps students see the dimensions of a rectangle as the factors of multiplication problems. Using their intuitive understanding of the distributive property and place value understanding, students begin to decompose one or both factors and partition open area models into proportional parts.

Study Margo's solution in Figure 4.24. Notice that her solution shows evidence of using place value understanding to decompose one of the factors. Most importantly, the solution shows evidence that multiplying each addend of one factor by the second factor and then adding the two products $(8 \times 10) + (8 \times 2)$ results in a product that is equal to the product of the original factors (8×12). As students work with larger numbers using open area models, they bring their place value understanding together with a developing understanding of the distributive property. Over time students no longer need to use the open area model, but apply the property inherent in the model to solve the problem using the distributive property, partial products, and the traditional US algorithm.

Go To Chapter 8: Understanding Algorithms has an in-depth discussion on how the open area model, place value, and understanding of the distributive property of multiplication come together to bring meaning to the partial products and traditional algorithms for multiplication and division.

Figure 4.24 Margo's response. Margo used an open area model to multiply 12 × 8.

John bought 12 boxes of crayons. Each box contained 8 crayons. How many crayons were there altogether?

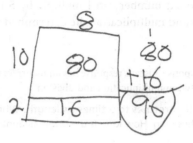

The Inverse Relationship Between Multiplication and Division

Just as every subtraction problem can be written as an addition problem, every division problem can be rewritten as a multiplication problem. For example, the division problem 12 ÷ 4 = ? can be rewritten as ? × 4 = 12 and interpreted as "What number multiplied by 4 equals 12?" In this way division can be thought of as a missing-factor multiplication problem. This highlights the inverse relationship between multiplication and division. "Understanding this relationship between multiplication and division is critical for learning division number facts and for dealing flexibly with problem situations involving multiplication and division" (Carpenter et al., 2003, p. 126).

The area models in Figure 4.25 provide an example of this inverse relationship and show that 7 × 10 = 70, 70 ÷ 7 = 10, and 70 ÷10 = 7.

Figure 4.25 Example using area models to show the inverse relationship between multiplication and division.

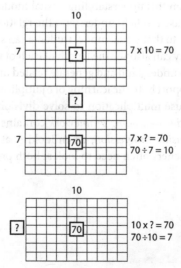

Students who understand the inverse relationship between multiplication and division can use multiplication to solve division problems. Study Oscar's response in Figure 4.26. It appears Oscar interpreted this division situation, 65 inches ÷ 5, as "What number can I multiply by 5 to equal 65 inches?" Through estimation and multiplication, he determined that $13 \times 5 = 65$.

Figure 4.26 Oscar's response. Oscar's response has an incorrect unit but is based on the inverse relationship between multiplication and division.

A piece of elastic stretches to 5 times its length. When fully stretched the elastic is 65 inches long. How long is the original length of the elastic?

$$\begin{array}{ccc} 5 & 5 & 5 \\ \times 11 & \times 12 & \times 13 \\ \hline 55 & 60 & 65 \end{array}$$

13 ft is the original length of the elastic

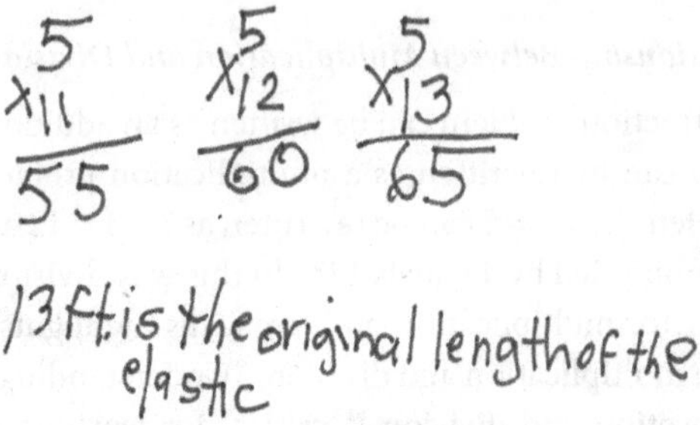

There are a number of ways to help students understand and become comfortable with the inverse relationship between multiplication and division. First, the idea of an inverse relationship between operations is not new to students. In grades 1 and 2 students use visual models and create related addition and subtraction equations to understand that subtraction is the inverse operation of addition. Building on this understanding, visual models like the area model in Figure 4.25 can be used to help understand the relationship between multiplication and division. In that visual model students can see the total number of square units is 70. They also see the length of each of the dimensions. Using models like this one to understand and generate related multiplication and division equations can support both the learning of multiplication and division facts and the flexibility to use multiplication to solve division problems. Although the *OGAP Multiplicative Reasoning Framework* contains both a *Multiplication Progression* and a *Division Progression*, when looking at student solutions for division problems, teachers often need to look at both progressions.

See Chapter 9: Developing Math Fact Fluency for a discussion on how to use understanding of the inverse relationship between multiplication and division to help students learn their division facts.

Chapter Summary

- Unitizing is a key understanding to help students move away from additive strategies to more sophisticated strategies for solving multiplication and division problems.
- Understanding and using place value and the properties of operations can help students develop procedural fluency with understanding to solve multiplication and division problems.
- Having students share and justify their strategies for solving multiplication and division problems is a context from which properties of operations can be highlighted, explored, and justified.
- Visual area models are foundational for understanding the properties of operations and other important multiplication relationships.

Looking Back

1. **Unitizing and Student Solutions**: As described in this chapter, unitizing underlies the concepts of multiplication, division, and place value. The following problem allows students the opportunity to simultaneously see that each octopus is both 1 group and 8 legs. This is an example of unitizing.

The Octopus Problem

Each octopus has 8 legs.
There are 6 octopuses at the aquarium.
How many legs are there in all?

Four student solutions to the octopus problem are shown. Examine each piece of work and identify how the students represented each octopus in their solution.

Figure 4.27 Four student solutions to the octopus problem.

Clarissa's Solution

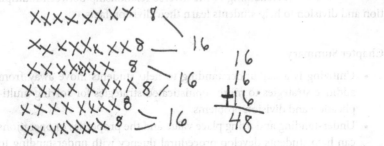

Carmi's Solution

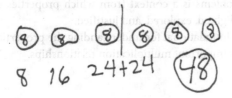

8   16   24+24   48

Shantel's Solution

$$
\begin{array}{r}
8 \\
\times\ 6 \\
\hline
48 \text{ legs}
\end{array}
$$

Maurice's Solution

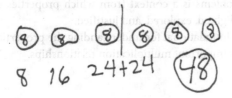

2. **Unitizing and Multiplication on an Array**: Students' ability to see different units in multiplicative situations can support fluency and flexibility when solving multiplication and division problems. Study the array of 24 objects shown in Figure 4.28. Identify the various units that comprise this array and indicate where you see each unit.

Figure 4.28 An array representing 24 objects.

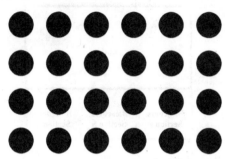

3. **The Commutative and Associative Properties of Operations**: Recall that the properties of operations "form the building blocks of all arithmetic" (Beckmann, 2014). The commutative and associative properties play a particularly important role in a student's understanding and fluent use of multiplication.

   (a) We learned in this chapter that the commutative property for multiplication states that the order in which two factors are multiplied does not affect the product. How is this property helpful as students are learning the basic multiplication facts?

   (b) The following two expressions represent the total number of objects in the array shown in Figure 4.28.
   - $(4 \times 3) \times 2$
   - $4 \times (3 \times 2)$

   Indicate on the array how each of these expressions represents the total number of objects in the array.

4. Study Juan's solution to the gymnasium problem shown in Figure 4.29. What property or relationship is evidenced in Juan's solution?

Figure 4.29 The gymnasium problem and Juan's solution.

The Gymnasium Problem

A sketch of the state college gymnasium is shown below.

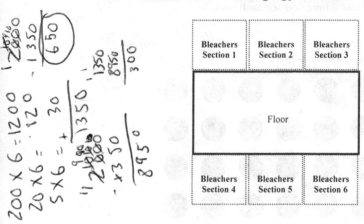

Each section of bleachers seats 225 people.

How many extra chairs need to be set up on the floor so that the gymnasium can seat 2000 people? Show your work.

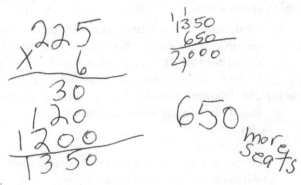

5. Devon's solution to the equation problem is shown in Figure 4.30. Identify the property or relationship evidenced in Devon's solution.

Figure 4.30 The equation problem and Devon's solution.

Write an equation to match this picture.

Explain your thinking.

6. **Place Value:** Using place value understanding and properties of operations, describe different ways that students can solve the following problems.
   (a) $3 \times 25 \times 4 =$
   (b) $5 \times 700 =$
   (c) $500 \times 400 \times 6000 =$
   (d) $45 \times 7 =$

7. **Multiplication by Powers of 10:** The rule, "When you multiply a number by 10, just add a zero to the end of the number," has too often dominated instruction related to multiplication by powers of 10. In this chapter you learned the inherent problems with this rule. Provide a sequence of five or more problems that can help students develop understanding of the

impact of multiplying or dividing by powers of 10. Indicate how you would engage students in these problems.

Instructional Link

Use the following questions to help you think about the ways your instruction and math program provide students opportunities to develop fluency with understanding through instructional emphasis on properties and other foundational concepts of multiplication.

1. To what degree do you or your math problem focus on place value understanding to develop fluency?
2. To what degree does your math instruction and program provide students intentional and systematic opportunities to understand and use the properties of operations to develop understanding of multiplication?
3. What modifications can you make to your instruction to ensure that students consistently engage with place value understanding and properties of operations as they are developing understanding and fluency with multiplication?

Resources to delve deeper into understanding of the properties of operations and place value are:

Battista, M. (2012). *Cognitively-based assessment of place value.* Portsmouth, NH: Heinemann.

Carpenter, T., Franke, M., & Levi, L. (2003). *Thinking mathematically: Integrating arithmetic & algebra in the elementary school.* Portsmouth, NH: Heinemann.

Empson, S., & Levi, L. (2011). *Extending children's mathematics: Fractions and decimals.* Portsmouth, NH: Heinemann. (Chapters 4 and 5).

Problem Contexts

Big Ideas

- The context of multiplication and division problems often influences the difficulty of problems and the strategy that students use to solve problems.
- Semantic structures provide a lens through which to understand the structural differences between the different contexts in multiplication and division problems.
- Reliance on immature strategies for solving multiplication and division problems can be offset by varying the context of problems.

Contexts

Equal groups
Equal measures
Measure conversions
Multiplicative comparisons
Patterns
Unit Rate
Rectangular area
Volume

This chapter illustrates the range of problem contexts students encounter while developing understanding and fluency when multiplying and dividing. As you read this chapter we suggest you have the *OGAP Multiplicative Reasoning Framework* on hand—particularly page 4, which has examples of different problem contexts that are discussed throughout the chapter. On page 1 of the framework notice the list of the contexts as shown above. These are the contexts that will be addressed in this chapter and are part of the CCSSM expectations for grades 2–6.

Note: This chapter uses the term *problem contexts* to distinguish between equal groups, equal measure, measurement conversions, multiplicative comparison, patterns, unit rates, area, and volume problems. This is not to be confused with contextual situations that describe the scenarios in which problems are found (e.g., sharing brownies compared to sharing marbles).

CCSSM and Multiplication and Division Contexts

Table 5.1 summarizes the multiplication and division contexts that students encounter as they develop the understanding, flexibility, and fluency that are expected in the CCSSM. Importantly, one can see the progression of problem contexts across the grades starting with equal groups in grade 2 to unit rates and constant speed at grade 6. The bold indicates problem contexts that are new to the grade.

Table 5.1 Summary of CCSSM context expectations for multiplication and division of whole numbers for grades 2–6.

Grade	CCSSM Multiplicative Problem Contexts (BOLD indicates new for that grade level)
2	**Equal groups**
3	Equal groups, **arrays, equal measures, introduction to area**
4	Equal groups, equal measures, **multiplicative comparisons, measurement conversions within systems, area, patterns**
5	Equal groups, equal measures, multiplicative comparisons, area, **measurement conversions between systems, the concept of volume, patterns, and scaling**
6	Volume, **unit rate including those involving unit pricing and constant speed, and common factors and multiples**

The focus of this chapter is on how the context of multiplication and division problems influences the difficulty of problems. To understand and describe distinctions between the different multiplication and division contexts, researchers use the term *semantic structures*. The semantic structures that influence the difficulty of problems are associated with the quantities in the problem and how they relate to each other (Bell, Fischbein, & Greer, 1984; Bell, Greer, Grimison, & Mangan, 1989; Brown, 1982; Carraher, Carraher, & Schlieman, 1987; De Corte, Verschaffel, & Van Coillie, 1988; Fischbein, Deri, Nello, & Marino, 1985; Nesher, 1988; Vergnaud, 1988).

Specifically, semantic structures in problems include:

1. The types of quantities in the problems (e.g., equal groups, dimensions, scale factors, conversion factors)

2. How the quantities interact with each other in the problem (e.g., multiple groups, repeated measure, change of size, conversion to new unit)
3. Impact on units (e.g., new unit created, scaling up or down, conversions)

Unlike other problem structures that are discussed in Chapter 6 (e.g., magnitude of factor, number of factors) that can be applied to problems of any context (e.g., How many wheels in ___ tricycles?), semantic structures are specific to the context of the problem.

Multiplication and Division Contexts

This chapter includes a discussion about different problem contexts and their semantic structures for equal groups, equal measures, measurement conversion, multiplicative comparisons, multiplicative patterns, area, volume, and unit rates. At the end of the chapter you will be asked to reflect on the degree to which your instruction or your math program varies the problem contexts consistent with the demands of the grade level shown in Table 5.1. As you read through the rest of the chapter, think about the implications that these different contexts and their unique semantic structures have on your instruction. At the end of the chapter Table 5.4 provides a summary of the discussion about each problem context.

Equal Groups

Equal groups problems involve iterations of many-to-one relationships as discussed in Chapter 1. The basic structure of equal groups problems involves multiple groups (Bell et al., 1989). For example, in the following problem there are 68 groups with 4 wheels in each group.

There are 68 cars in the parking lot. Each car has 4 wheels.
How many wheels are there in all?

The quantities in the problem are the number of cars, and the composite unit is the number of wheels on each car. The solution is expressed as the total number of wheels. Look closely at the equation that represents this problem situation:

68 cars × 4 wheels on each car = 272 wheels

Notice that the unit associated with each quantity is different. Particularly note that the answer (total number of wheels) involves a different unit than either the multiplier (number of cars) or the multiplicand (wheels on each car). Creating a new unit when multiplying or dividing is a new idea for students when they first engage in multiplication and division (Schwartz, 1988; E. Silver, personal communication, 2006). Although one can fashion an addition problem with different units such as 5 cars + 4 trucks = 9 vehicles, the general convention in addition and subtraction problems is that the quantities in the problem describe the same unit, unlike in multiplication and division.

When students first solve equal groups multiplication problems, they often use additive strategies. That is, students recognize the equal groups and then either count by ones or use repeated addition as evidenced in the student work samples in Figures 5.1 and 5.2, respectively.

Notice in Figure 5.1 that Samantha drew each window pane. Pencil marks in each of the panes of the windows suggest she counted each pane to determine the number of panes in 4 windows. On the *OGAP Multiplication Progression* this is considered an *Early Additive Strategy*.

Figure 5.1 Samantha's response. The tick marks on Samantha's response provides evidence of a counting by ones strategy when solving this problem.

A classroom has 4 windows. Each window is divided into small panes like in the picture. How many panes of glass are there in the 4 windows in the classroom?

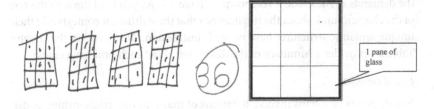

Figure 5.2 Wyatt's response. Wyatt used repeated addition and a building up strategy to solve this equal groups multiplication problem.

Mark bought 12 boxes of crayons. Each box contained 8 crayons. How many crayons are there in all?

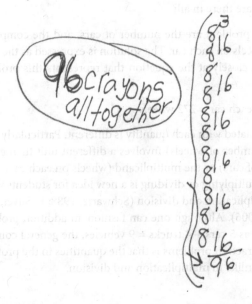

In contrast, Wyatt, in Figure 5.2, does not need to represent each crayon and instead operates with the composite unit (8 crayons in a box). Wyatt's solution also shows evidence of building up, which is an *Early Transitional Strategy*.

Equal Measures

Equal measures problems also involve many-to-one situations. The basic structure of equal measures problems involves repeated or iterated measures of a continuous quantity (Bell et al., 1989). For example, in the problem and equation that follow the repeated measure is 4 feet of ribbon in each bow. The multiplier is 14 bows. The solution is the total feet of ribbon.

> One bow uses 4 feet of ribbon. How many feet of ribbon are needed to make 14 bows?

The quantities—number of bows, feet of ribbon in each bow, and total feet of ribbon—are represented in the following equation:

14 bows × 4 feet of ribbon in each bow = 56 feet of ribbon

A related equal measures division problem is shown here:

> Tyler is making bows that each require 4 feet of ribbon.
> How many bows can Tyler make with 56 feet of ribbon?

Notice that the quantities (14 bows, 4 feet of ribbon per bow, and 56 feet of ribbon) are used in both problems. The multiplicative relationship among these quantities is also the same in both problems. The difference is the unknown quantity contained in each problem. In the original problem the total number of feet needed is the unknown, and in the second problem, the number of bows is the unknown quantity. This small change makes the second problem a division or a missing factor problem. The following two equations illustrate these two interpretations of this problem:

> Division interpretation: 56 feet of ribbon ÷ 4 feet in each bow = n bows
> Missing factor interpretation: n bows × 4 feet in each bow = 56 feet of ribbon

This is an example of the inverse relationship between multiplication and division.

 Go To For a more detailed discussion of division see Chapter 7: Division.

Unit Rates

Unit rates become an instructional focus in the CCSSM starting at grade 6. Unit rates play an important role in bridging elementary school multiplication and division contexts with proportionality in the middle grades. As has been stated many times throughout this book, by the end of fifth grade students should have a deep understanding of, and flexible and efficient methods for, solving problems involving multiplicative relationships in a range of contexts and contextual situations in order to be prepared to engage in middle school topics. Proportionality, in particular unit rates, is often the first of these middle school topics.

Unit rate problems are many-to-one situations that involve a special ratio that compares two quantities with different units of measure. As previously indicated, equal groups and equal measures problems also involve many-to-one situations (e.g., balloons in a bunch or feet per bow). One thing that distinguishes rate problems from equal groups and equal measures problems, however, is that rate problems involve recognizable quantities (e.g., price per pound, miles per hour, miles per gallon) rather than the arbitrary quantities created in equal group and equal measure problems. That is, a rate is a more commonly accepted quantity.

Although the CCSSM does not expect students to formally engage in rate problems until grade 6, students in earlier grades encounter unit rates in their everyday lives. From that exposure students will begin to build an under-standing of the relationships in rates (e.g., if apples cost $2.00 per pound, then 3 pounds of apples will cost $6.00). As students enter middle school teachers can draw on these earlier experiences to work with more complicated and less intuitive rates like speed.

Speed is one of the "most common rates" and at the same time "poorly understood by most people" (Lamon, 2005, p. 203). Researchers have found that students have a difficult time understanding that speed cannot be mea-sured directly; rather it is a measure of motion that comes about by comparing two other quantities (the ratio of distance to time) (Lamon, 2005).

In the case of speed as a rate, the two different quantities can be miles and hours. Considered independently, miles define distance, and hours define time. As a rate they form a new unit that describes the relationship between miles and time. This new unit is a measurement of speed.

Compare Margo's and Alex's responses in Figure 5.3. What is the evidence in Margo's response that she understood the relationship between distance and time in the problem? What is the evidence in Alex's response that makes you question his understanding of the relationship between distance and time in the problem?

As you probably noticed, Margo's solution shows strong evidence of under-standing the relationships between distance and time represented in the speed of 8 miles per hour. Margo identified the distance traveled after 1 hour and then 2 hours using the multiplicative relationship. Additionally, she recognized 4 miles would take half the time as 8 miles. In contrast, there is no evidence that Alex understood the relationship between distance and time in the problem. Alex divided 20 by 8, ignored the remainder, and decided that the quotient was 2 miles per hour.

Figure 5.3 Margo's and Alex's responses. The evidence in Margo's and Alex's responses show a very different level of understanding of the meaning of 8 miles per hour.

> Bob rides his bike at a speed of 8 miles per hour. How long did it take him to ride 20 miles? Show your work.

Margo's Response

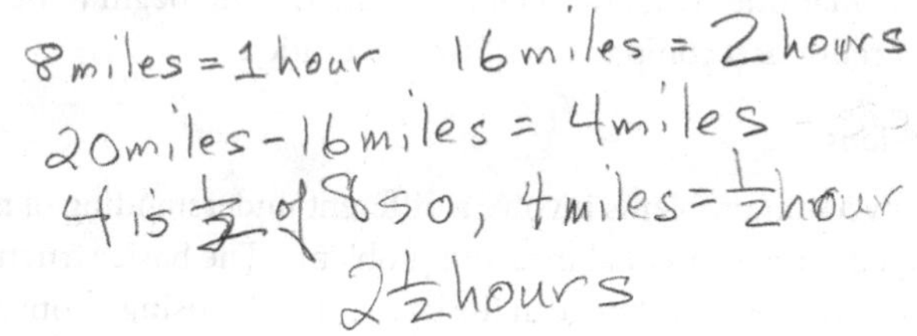

$$8 \text{ miles} = 1 \text{ hour} \quad 16 \text{ miles} = 2 \text{ hours}$$
$$20 \text{ miles} - 16 \text{ miles} = 4 \text{ miles}$$
$$4 \text{ is } \tfrac{1}{2} \text{ of } 8 \text{ so, } 4 \text{ miles} = \tfrac{1}{2} \text{ hour}$$
$$2\tfrac{1}{2} \text{ hours}$$

Alex's Response

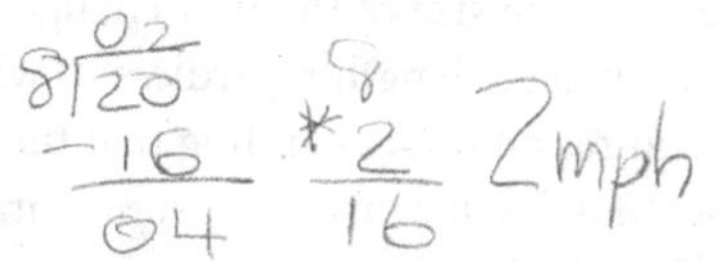

$$8 \overline{)20} \quad \ast 2 \quad 2 \text{ mph}$$
$$\begin{array}{r} 02 \\ -16 \\ \hline 04 \end{array} \quad \begin{array}{r} 8 \\ \ast 2 \\ \hline 16 \end{array}$$

One strategy to help students like Alex build an understanding of speed as a measure that comes from comparing the ratio of distance to time is to introduce a double number line (see Figure 5.4) while asking questions focused on the relationship between distance and time. Study the double number line in Figure 5.4. Notice the two parallel number lines coordinate the relationship between distance traveled and time.

Figure 5.4 A double number line illustrating the relationship between miles and hours as miles per hour.

> Bob rides his bike at a speed of 5 miles per hour. How far will Bob travel if he rides his bike at the same speed for 3 hours?

At the rate of 5 miles per hour Bob rides his bike 15 miles in 3 hours

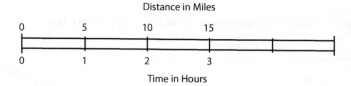

Distance in Miles

Time in Hours

Sample Questions:

1. What does the double number line tell you about how far Bob travels after 1 hour if he is riding his bike at 5 miles per hour? Two hours? How do you know?

2. If Bob continues to ride his bike at 5 miles per hour, how far will Bob travel in 4 hours? How do you know?

3. How much time will it take Bob to travel 25 miles? How do you know?

By having students interact with a range of unit rate problems (e.g., speed, unit pricing), using double number lines or other strategies that show the relationships between the quantities (e.g., ratio tables), students will begin to build an understanding of ratios as a comparison of two quantities.

Measure Conversions

Measurement conversion problems involve a different understanding of multiplication than equal groups or equal measure problems. The basic structure of conversion problems involves a change in the size of the unit using a conversion factor (Bell et al., 1989). The conversion factor is a many-to-one relationship (e.g., 12 inches to a foot). In measurement conversion problems the total amount does not increase or decrease; only the size of the unit changes. That is, a single piece of ribbon has the same absolute length regardless of whether it is measured in inches or feet. This requires an understanding that the larger the unit of a measure, the fewer number of units in a given measure (Sarama & Clements, 2009). For example, it takes more inches than feet to measure a given distance because inches are smaller units than feet (see Figure 5.5).

In the following problem, the conversion factor is 1 foot = 12 inches. To be successful students need to understand the relationship between these different size units. That is, there are 12 inches in every foot.

A bow uses 2 feet of ribbon. How many inches of ribbon are needed to make the bow? (1 foot = 12 inches)

2 feet of ribbon × 12 inches for each foot = 24 inches

Figure 5.5 illustrates how the quantities in this problem are related. For every 1 foot of ribbon there are 12 inches of ribbon. Therefore, 2 feet of ribbon is equal to 24 inches of ribbon. The size of the measurement unit and the number of units change, but the length of the ribbon stays the same.

Figure 5.5 Visual model illustrating 2 feet of ribbon equal to 24 inches.

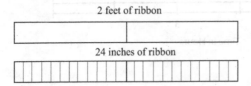

2 feet of ribbon

24 inches of ribbon

Figure 5.6 is an example of a measurement conversion problem that involves converting ounces to pounds. Tyler applies the conversion factor accurately to solve the problem.

Figure 5.6 Tyler's response. The evidence in his work suggests the number of ounces was correctly converted to pounds.

> Katrina carried some groceries home from the deli. Here is what was in her bag.
> 36 ounces of cheese
> 16 ounces of turkey
> 16 ounces of ham
> 4 ounces of roast beef
> How many pounds of food did she carry? Show your work.
> 1 pound = 16 ounces

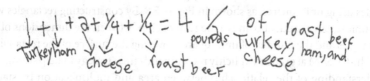

Figure 5.7 is a multistep problem in which students first interact with an equal measures problem (20 cans each containing 355 ml of soda) and then convert milliliters of soda to liters of soda. Gavin accurately found the total number of liters in 20 cans of soda and then accurately converted milliliters to liters of soda.

Figure 5.7 Gavin's response. Gavin accurately solved this multistep problem involving equal measures and a conversion by first determining the total number of liters in 20 cans of soda and then converting milliliters to liters.

> Trina bought 20 cans of soda. Each can of soda contains 355 milliliters. How many liters of soda did she buy? [1000 milliliters = 1 liter]

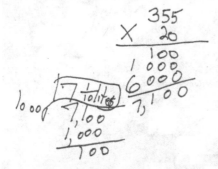

In an OGAP pilot of this item, 14 out of 26 (53 percent) sixth grade students solved this question correctly. The remaining 47 percent of the students attempted or correctly solved the equal measures portion of this question but not the conversion portion. This may be an example in which the change in semantic structure from an equal measures to a conversion factor influenced student solutions.

Area and Volume

An important semantic feature of area and volume problems is that they involve dimensions. More specifically, area problems require an understanding that the product of the two dimensions of a rectangle identifies the number of unit squares that completely cover the area of the rectangle. Likewise, inherent to volume problems is the notion that multiplication of the three dimensions of a rectangular prism counts the total number of unit cubes that completely fill the prism. As with other multiplication situations discussed previously, the product in area and volume problems describes a different unit than the associated factors. The product in area problems is a square unit, and the product in a volume problem is a cubic unit. To build understanding of these new units and the relationship of area to multiplication, students should engage in problems in which they build rectangles using unit squares, as shown in Figure 5.8. By constructing rectangles with the unit squares, students can see the relationship between the dimensions of the rectangles and the number of unit squares covering the surface of the rectangle.

The CCSSM at grade 3 explicitly recommends this strategy to help strengthen understanding of the relationship between area and multiplication by stating that students are to "find the area of a rectangle with whole number side lengths by tiling it, and show that the area is the same as would be found by multiplying the side lengths" (CCSSO, 2010).

Figure 5.8 Using unit squares to build understanding of area as the product of dimensions. In this case 15 square units is the product of 3 units × 5 units.

Problems involving area and volume first draw on the understandings developed using squares and cubes, respectively, as shown in Shawna's work in Figure 5.9 and Thomas's work in Figure 5.10.

Figure 5.9 Shawna's response. Shawna determined the number of unit squares that comprise a rectangle with 6 rows and 5 unit tiles in each row.

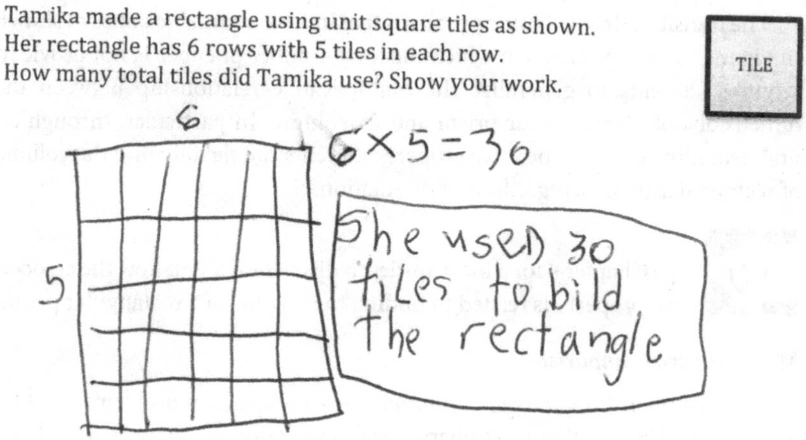

Tamika made a rectangle using unit square tiles as shown. Her rectangle has 6 rows with 5 tiles in each row. How many total tiles did Tamika use? Show your work.

TILE

6

5

$6 \times 5 = 30$

She used 30 tiles to bild the rectangle

Figure 5.10 Thomas's response. Thomas decomposed the original figure to show that the volume is the same in Figure 1 as in Figure C.

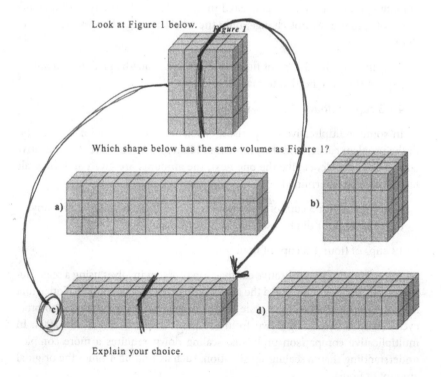

Look at Figure 1 below.

Figure 1

Which shape below has the same volume as Figure 1?

a)

b)

c)

d)

Explain your choice.

C - IF you cut it likE I cut it - it is the same as C.

The transition from volume problems that involve countable cubes as shown in Figure 5.10 to problems in which the total number of cubes is not provided requires students to generalize the multiplicative relationship between the dimensions of a rectangular prism and its volume. In particular, through an understanding of the associative property students can flexibly find the volume of rectangular prisms, regardless of orientation.

 See Chapter 4 for a more in-depth discussion about how the associative property is related to finding the volume of a rectangular prism.

Multiplicative Comparisons

The key distinguishing feature of multiplicative comparison problems involves scale factors. Unlike other multipliers, a scale factor does not have an associated unit. The scale factor indicates how many times more or how many times less one quantity is than another quantity.

For example, in the problem that follows the multiplier—4 times the original amount—does not have an associated unit. Additionally, the type of quantity (cups of flour) does not change; only the total amount changes (12 cups of flour).

A recipe requires 3 cups of flour. If the recipe is quadrupled, how many cups of flour are needed to make the recipe?

4 × 3 cups of flour = 12 cups of flour

In some multiplicative comparison problems, the scale factor is given (as in the problem earlier and the problem in Figure 5.12). In other multiplicative comparison problems, like the one next, the amounts are given and the scale factor must be determined.

A recipe calls for 3 cups of flour. Robert used 12 cups of flour. How many times more flour did he use than in the original recipe?

12 cups of flour ÷ 3 cups of flour = 4

A third type of multiplicative comparison problem involves using a scale factor to scale down a value. Read the motorcycle problem that follows. Notice that the solution involves using a scale factor to determine the number of motorcycles in 1960 by scaling down from the number of motorcycles in 2004. In multiplicative comparison problems, scaling down requires a more complex understanding than a scaling-up situation, such as finding 4 times the original amount of flour.

In 2004 there were about 5,760,000 motorcycles in the United States. That is about 10 times more than the number of motorcycles in 1960. About how many motorcycles were there in the United States in 1960?

Examine the three solutions in Figures 5.11–5.13 to this problem involving scaling down. What do you notice about how the students solved the problem?

Figure 5.11 Leah's response. Leah disregarded the meaning of the quantities in this problem and inappropriately subtracted the scale factor from the number of motorcycles.

$$
\begin{array}{r}
5,7\overset{5\,9\,9\,1}{6\emptyset,\emptyset00} \\
-\qquad 10 \\
\hline
5,759990
\end{array}
$$

There are about 5,759990 in the United States

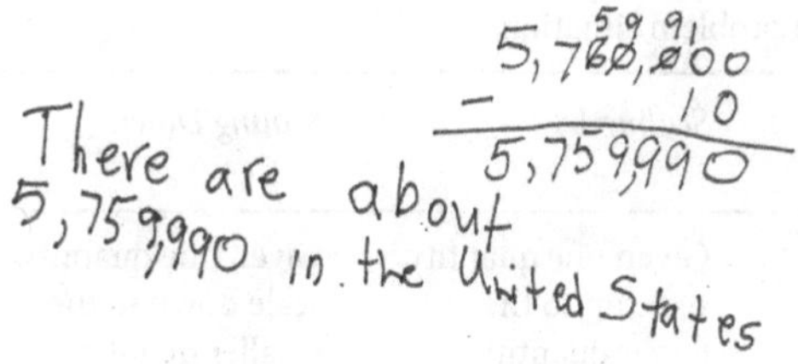

Figure 5.12 Luke's response. Luke correctly interpreted this multiplicative comparison situation as evidenced when he used the scale factor to determine the number of motorcycles in 1960.

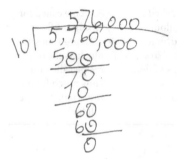

Figure 5.13 Thomas's response. Thomas incorrectly found 10 times the number of motorcycles in 2004.

$$
\begin{array}{r}
5,760000 \\
\times\qquad 10 \\
\hline
57,600,000
\end{array}
$$

57,600,000 moter cycles

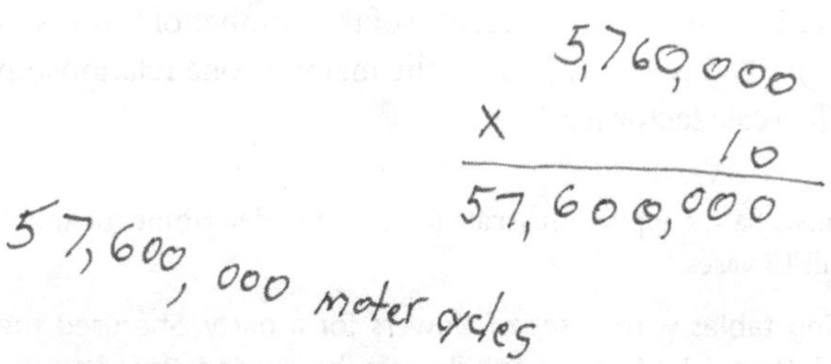

You probably noticed that only Luke determined the number of motorcycles in 1960 by dividing the number of motorcycles in 2004 by 10. Leah incorrectly subtracted 10 from the number of motorcycles in 2004, and Thomas incorrectly multiplied the number of motorcycles in 2004 by 10. Solutions like Leah's and Thomas's are not uncommon when students are asked to solve problems involving scaling down, which is more complex than scaling up. In a small OGAP study of 51 students, 35 percent of the students solved this problem like either Luke or Leah.

Table 5.2 summarizes the differences between the different types of multiplicative comparison problems exemplified in this section. Note that the equations represent the problem situation but do not necessarily represent the strategy that students will use to solve the problem.

Table 5.2 Multiplicative comparison problem situations.

	Finding the Scale Factor	Scaling Up	Scaling Down
Situation	Given two quantities, find the scale factor between the two quantities	Given one quantity, scale up to the larger quantity	Given one quantity, scale down to the smaller quantity
Equation (n is the unknown)	Larger quantity ÷ smaller quantity = n (scale factor)	Scale factor × smaller quantity = n (larger quantity)	Quantity ÷ scale factor = n (smaller quantity)
Example	Sam used 12 cups of flour in a recipe. The original recipe called for 3 cups of flour. How many times more flour did he use then was required in the original recipe?	A recipe called for 3 cups of flour. Sam quadrupled the recipe. How many cups of flour did he use after quadrupling the recipe?	A recipe called for 12 cups of flour. That is 4 times as much as the original recipe called for. How much flour did the original recipe call for?

Multiplicative Patterns

Multiplicative patterns involve the application of a scale factor in many-to-one situations. To understand how the scale factor applies, study the problem and Saffra's solution in Figure 5.14. Notice that the ratio of the number of flowers to the number of vases to be filled is 4 to 1. That is, the many-to-one relationship is 4 flowers to 1 vase. The scale factor is × 4.

Figure 5.14 Saffra's response. Saffra applied the scale factor × 4 to determine the number of flowers needed to fill 15 vases.

Tammy is decorating tables with vases of flowers for a party. She used the following chart to keep track of how many flowers she needed. Based on the information in the table, how many flowers does she need to fill 15 vases?

Number of Vases of Flowers	Number of Flowers
1	4
2	8
3	12
4	16
15	60

CCSSM and Multiplicative Patterns

The CCSSM emphasis at grade 4 is on students examining, describing, and extending number and shape patterns. At grade 5 this work is extended to studying the relationships between ordered pairs in tables and graphs, as well as graphs of ordered pairs on the first quadrant of a coordinate plane. This work lays the foundation for studying proportional relationships and functions in middle school. Of note the CCSSM emphasis on patterns "does not require students to infer or guess the underlying rule for a pattern, but rather asks them to generate a pattern from a given rule and identify features of the given pattern" (Common Core Standards Writing Team, 2011).

Marco's solution in Figure 5.15 is an example of generating a pattern from a given rule and identifying the features of a pattern. In this case Marco identified the multiplicative relationship (doubling) between the two variables or ordered pairs.

Figure 5.15 Marco's response. Marco identified and applied the multiplicative scale factor (doubling) between ordered pairs.

Complete the table following the rules listed in the column headings. What do you notice about how the numbers in the two columns relate? Explain why.

Add 2	Add 4
2	
4	8
6	12
8	16
10	20
12	24

The number in the first column is always doubled. The number in the last column is always half of itself in the first column.

Because it gave me the 4 and come out as an 8. The same with 6 and 12. So that tells me to double the number I have to fill in.

Not all students recognize or describe the multiplicative relationship between the ordered pairs, as can be seen in Albert's solution in Figure 5.16.

Figure 5.16 Albert's response. Albert identified an additive rather than a multiplicative relationship between the quantities in the ordered pairs.

Complete the table following the rules listed in the column headings. What do you notice about how the numbers in the two columns relate? Explain why.

Add 2	Add 4
2	4
4	8
6	12
8	16
10	20
12	24

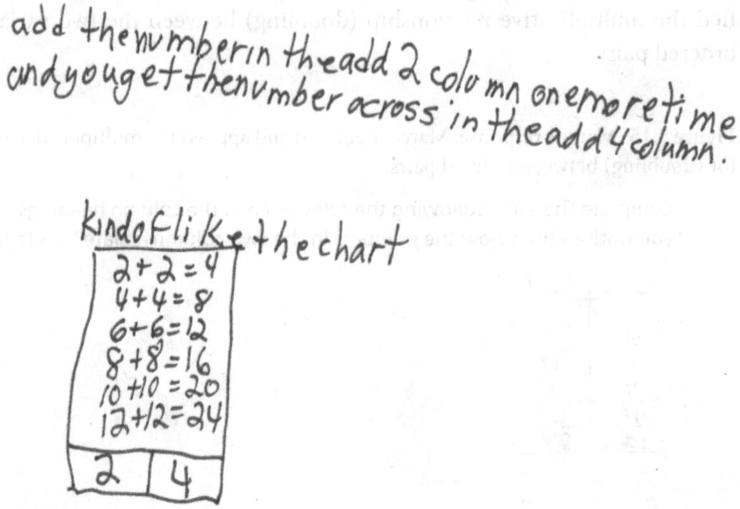

add the number in the add 2 column and you get the number across in the add 4 column one more time.

kind of like the chart

2+2=4
4+4=8
6+6=12
8+8=16
10+10=20
12+12=24

2	4

Instructionally, this is a good opportunity to show both solutions to engage students in a focused discussion that helps students recognize the multiplicative relationships.

1. How are the solutions alike? Different?
2. Generate other patterns that have doubling patterns (or tripling patterns, and so on).
3. If you know the number in column one, can you determine the number that will be in column two? How?

The next section provides a case study that reinforces the impact of students interacting with different problem structures.

Impact of Semantic Structures: Instructional Implications

To understand the potential impact of the different semantic structures found in different contexts researchers study student responses to arithmetically equivalent problems that are semantically different (De Corte et al., 1988). To explore the impact of semantic structures on the challenge of different problem situations, solve the problems in Task 1 and Task 2. How are these problems alike and how are they different?

> **Task 1:** The school band stores instruments in a closet that has 13 shelves in it. There are 117 instruments in the closet. Each shelf holds the same number of instruments. How many instruments are on each shelf?
>
> **Task 2:** The typical house mouse can run at a top speed of 13 km per hour. The typical cheetah can run at a top speed of 117 km per hour. How many times faster is the cheetah than the house mouse when they are each running at top speed?

Notice these problems are arithmetically equivalent but semantically different. That is, both Task 1 and Task 2 involve the division of 117 by 13 (arithmetically equivalent). However, because the problems vary semantically, the tasks have very different levels of difficulty. Study Shelby's solution to both problems in Figures 5.17 and 5.18.

Figure 5.17 Shelby's response to Task 1. Shelby correctly determined the number of instruments on each shelf.

Task 1: The school band stores instruments in a closet that has 13 shelves in it. There are 117 instruments in the closet. Each shelf holds the same number of instruments. How many instruments are on each shelf?

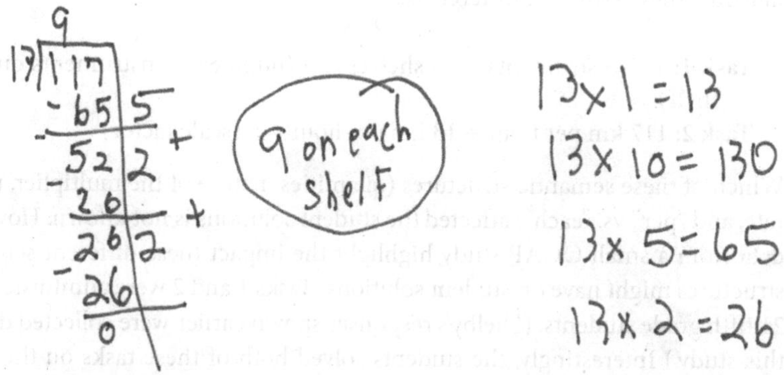

Figure 5.18 Shelby's response to Task 2. Shelby incorrectly interpreted the problem as multiplication.

> **Task 2:** The typical house mouse can run at a top speed of 13 km per hour. The typical cheetah can run at a top speed of 117 km per hour. How many times faster is the cheetah than the house mouse when they are each running at top speed?

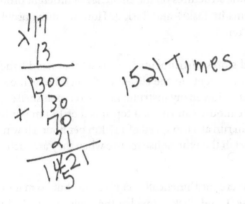

Semantically, these are different problems. Task 1 involves equal groups in which the total number of instruments (total) and the number of shelves (number of groups) is given. The solution involves determining the number of instruments on each shelf (number in each group). In contrast, Task 2 involves comparing two rates. The speed for each animal is given (13 km per hour and 117 km per hour). The unknown, or the quantity asked for in the problem, is the scale factor. Researchers indicate that both speed and scale factor are difficult concepts for students to understand (Lamon, 2005), and therefore add a level of challenge to Task 2 that is not found in equal groups problems such as Task 1.

The equations for Tasks 1 and 2, written here with associated units, help illustrate these semantic differences:

Task 1: 117 instruments ÷ 13 shelves = n (number of instruments on each shelf)

Task 2: 117 km per hour ÷ 13 km per hour = n (scale factor)

Which of these semantic structures (quantities, nature of the multiplier, use of rate, and "per" vs. "each") affected the student solutions is not known. However, data from a small OGAP study highlight the impact these different semantic structures might have on student solutions. Tasks 1 and 2 were administered to 21 fifth grade students. (Shelby's responses shown earlier were collected during this study.) Interestingly, the students solved both of these tasks on the same day; in fact both tasks were even presented on the same page. However, in all but four cases, students solved the two problems differently. Table 5.3 summarizes the differences in student responses to the two problems.

Table 5.3 Solution strategies for Task 1 and Task 2 for 21 fifth grade students.

Strategies	Number of Students (%) (n=21)
Division using an efficient procedure on both tasks.	4 (19%)
Division on both Task 1 and 2. However, the strategy for Task 2 was less efficient.	5 (20%)
Division at a range of sophistication for Task 1, but used multiplication for Task 2.	12 (60%)

As Table 5.3 shows, all 21 students interpreted Task 1 correctly as a division problem. However, five students used a less efficient strategy on Task 2, and over half of the students misinterpreted Task 2 as a multiplication problem (perhaps because of the word "times").

The results from this small classroom study are consistent with research that student strategies move from more to less sophisticated strategies as they are introduced to new problem structures (Kouba & Franklin, 1995; OGAP, 2006). Importantly, Kouba and Franklin (1995) have also found that it is important for students to interact with a range of contexts with different semantic structures to help overcome the use of immature strategies.

Together, these two pieces of research can be used to help make purposeful and knowledgeable instructional decisions. First, multiplication and division are found in a range of everyday problem contexts; elementary students should interact with the different contexts discussed in this chapter. Second, understanding that student strategies often move from more to less sophisticated strategies as they are introduced to problem structures (e.g., change of context, change of magnitude of factors) helps teachers be instructionally proactive.

Sample proactive or anticipatory instructional moves:

1. *Focused questioning using student solutions*: After students have solved a problem involving a new context, engage students in an understanding of the mathematics in the problem by asking focused questions about their solutions. Looking back at Shelby's solution in Figure 5.18 you might ask Shelby and other students with similar solutions questions like: A) What does it mean when it says the house mouse runs 13 km per hour? Sketch a picture. B) What does it mean when it says the cheetah runs 113 km per hour? Sketch a picture. C) Which animal runs farther after 1 hour? Explain how you know. D) Which animal runs the fastest? How do you know? E) About how much faster does the cheetah run than the house mouse? How do you know?

2. *Word problem strategy*: Use the word problem strategy described at the end of this chapter to focus students on understanding the problem situation as they are introduced to new problem contexts.

The *OGAP Multiplicative Reasoning Framework*: **Problem Contexts**

(OGAP) All problem contexts are listed on the front page of the *OGAP Multiplicative Reasoning Framework* under the heading Application/Context. You can use this knowledge about problem contexts to assure that your students interact with a range of problems with different semantic structures as appropriate for the grade level. Also remember the importance of varying other problem structures (discussed in Chapter 6) and of using problems that address foundational mathematics concepts and properties (discussed in Chapter 4).

Fifth and sixth grade students should fluently use strategies at the *Multiplicative* level for whole numbers, regardless of the problem context or other problem structures. Teachers using the OGAP formative assessment system have found it important to keep records of the problem contexts students have engaged in and the level on the progression that is evidenced in their work. Teachers have found that this information is important because it informs their instructional decisions. For example, if students are using *Multiplicative Strategies* for solving both equal measures problems and conversion problems, a teacher might decide to engage them in a multistep problem that involves two different problem contexts such as the problem in Figure 5.7.

Strategies to Help Students Solve Multiplication and Division Problems

Most teachers will say that students have difficulty solving word problems. This chapter, as well as Chapter 6: Problem Structures, provides explanations for why solving word problems are challenging for some students. Everything from the magnitude of factors to the semantic structures in different contexts influence the difficulty students have with word problems. The questions teachers always ask is: "How can we help students solve word problems?"

This section presents two different approaches to help students become more confident and competent solving word problems: 1) exposing students to a wider range of problem contexts and range of problem structures (Kouba & Franklin, 1995) and 2) adopting a literacy practice to engaging students in making sense of word problems.

Exposing Students to a Wider Range of Problem Situations

As has been stated, researchers have found that varying problem structures and contexts can help students develop flexibility and fluency when solving word problems (e.g., Kouba & Franklin, 1993). This approach is not necessarily intuitive; some teachers and some textbooks support the notion that more of the same type of problem is better than varying the problem contexts and structures. For example, teachers might think that until all students can solve equal group problems, new contexts shouldn't be introduced.

In contrast Kouba and Franklin (1993) suggest exposing students to a range of problem contexts in the classroom, as well as in other subjects and in daily life, rather than just during the designated time for mathematics. They recommend that students keep a journal about multiplicative situations they see in school; at home; in stores; and in newspapers, magazines, and the Web. Additionally, teachers should intentionally engage students in multiplication and division problems in different subject areas and different everyday applications consistent with the CCSSM demands at specified grade levels (see Table 5.1).

Adopting a Literacy Practice to Engage Students in Making Sense of Word Problems

Another strategy that has been adopted by OGAP teachers that they have found effective in helping students solve word problems was adapted from a research-based reading comprehension strategy: *read, retell, and anticipate next* (Gambrell, Koskinen, & Kapinus, 1991; Morrow, 1985). As you will see, this strategy puts the emphasis on understanding the context and the contextual situation before students solve the problem.

The strategy involves four parts that are exemplified here:

1. Remove the question from the problem and have students read and retell the situation (read and retell).
2. Have students generate questions that can be asked and answered given the problem situation (anticipate next).
3. Have students solve the problems that are generated.
4. Read the original question and have students solve it.

The following problem is used to illustrate these strategies.

During a physical education class 24 students played soccer, 8 students played basketball, and 16 students played kickball. Each student only played one game. What fraction of the students played soccer?

Step 1: Remove the question from the problem situation and have students retell the problem situation.

During a physical education class 24 students played soccer, 8 students played basketball, and 16 students played kickball. Each student only played one game.

When retelling the problem situation, have students turn to a partner and retell the story in their own words. Provide students about one to two minutes to do this. At the end of the two minutes ask someone to retell the story to the class. List the facts that were given in the story on chart paper (e.g., each student played only one game, 24 students played soccer, 8 students played basketball, 16 students played kickball).

Step 2: Have students generate questions.

Have students work with a partner to generate three to five questions that can be answered given the context and the information provided in the problem. Before you read further generate some questions that you think students might ask based on the situation and the information given in the problem.

You probably noticed that the context and the numbers in the problem allow for many more questions than the actual question associated with the problem. That is, the problem allows for addition (e.g., how many students altogether?), subtraction (e.g., how many more students played soccer than basketball?), division (e.g., how many times more students played soccer than basketball?), or fractions (e.g., what fraction of the students played basketball?).

After partners have completed their list of questions, have students share their questions with the class. Post all the questions for the full class to see. Students are usually very surprised by all the different questions that can be answered using similar information.

Step 3: Have students solve questions generated by the class that help meet the lesson goals.

This is a good opportunity to let students select the questions that they are interested in solving, or it can be a very good opportunity for you to differentiate instruction by assigning students different problems. Anticipating questions that students might generate before engaging students with the problem allows the teacher to think about which questions might be assigned to which students.

Step 4: Unveil the original question and have students solve the problem.

This word problem strategy should be applied multiple times in order for students to reap the benefits of it. You may want to begin math class a few days a week using this strategy. If students run into difficulty solving problems on their own, you can ask them to cover the question and retell the problem situation to refocus them. Students will get used to this strategy and understand the purpose.

The two strategies described here can help students see that multiplication and division involve a range of everyday problem situations, and it can help provide them with a strategy to engage in and develop flexibility when solving word problems.

The Limitations of Key Words as a Strategy

A strategy that many teachers use to help students solve word problems is to have students underline or highlight *key words* to help identify the operation embedded in the problem. However, researchers indicate that teaching the use of key words is a "limiting, detrimental strategy" (Kouba & Franklin, 1993, p. 106) for multiple reasons. First, key words often take on multiple meanings in word problems. Students are often taught that the word altogether implies addition. However, the following problem requires multiplication rather than addition to find the total amount:

> Sam had 3 boxes with 4 balloons in each.
> He bought twice as many balloons at the store.
> How many balloons does he have altogether?

Second, many word problems do not contain any key words, as shown in the following example:

> One bow uses 4 feet of ribbon.
> How many feet of ribbon are needed to make 14 bows?

Third, this strategy encourages students to focus on specific words, rather than the context within which those words are used in the problem. This chapter has focused on the fact that understanding problem contexts involves focusing on the meaning of the problem situation, including the quantities in the problem, how the quantities interact, and the units. Students need to first make sense of the situation in order to understand and use the appropriate operation and strategy. Rather than isolating specific words, the focus should be on making sense of the context.

Summary

- It is important that students experience a variety of multiplicative problem contexts (as appropriate for each grade level) so that they can develop flexibility and fluency.
- The semantic structure of multiplication and division problems varies across different problem contexts. These semantic structures influence the strategies students use to solve the problems, as well as their understanding of the multiplicative relationship in different problem situations. Table 5.4 summarizes some of the differences in semantic structures between different multiplicative problem contexts.
- Students can learn to be successful problem solvers by focusing on the meaning of the problem situation—including the quantities in the problem, how the quantities interact, and the units—using strategies such as the ones described in this chapter.

Table 5.4 Semantic structures of multiplication problems.

Problem Situation	Distinguishing Semantic Structures	Example Problem and Associated Equation	Impact of Operation on Quantities
Equal Groups	• Iterated groups	There are 68 cars in the parking lot. Each car has 4 wheels. How many wheels are there in all? 68 cars × 4 wheels on each car = 272 wheels	Repetition of equal groups of objects resulting in the total number of objects. The objects in each group are the same as the objects in the total (e.g., wheels).
Equal Measures	• Iterated measures	One bow uses 4 feet of ribbon. How many feet of ribbon are needed to make 14 bows? 14 bows × 4 feet of ribbon in each bow = 56 feet of ribbon	Repetition of measures resulting in a total length, area, volume. The continuous measure in each group is the same as the measure in the total (e.g., feet of ribbon).
Unit Rates	• A ratio composed of two different quantities that names a new quantity	Bob rides his bike 5 miles per hour. How far will Bob travel if he rides his bike at the same rate for 3 hours? 3 hours × 5 miles per hour = 15 miles	Each of the quantities in the problem has a different meaning, for example, time (hours) × speed (miles per hour) = distance (miles).
Conversions	• Conversion factor	A bow uses 2 feet of ribbon. How many inches of ribbon are needed to make the bow? (1 foot = 12 inches) 2 feet of ribbon × 12 inches for each foot = 24 inches	The magnitude of the measure does not change, but the size of the measurement unit changes.
Multiplicative Comparisons	• Scale factor that has no associated unit	A recipe requires 3 cups of flour. If the recipe is quadrupled, how many cups of flour are needed to make the recipe? 4 × 3 cups of flour = 12 cups of flour	The size or amount of a quantity scales up or down, but the type of quantity remains the same (e.g., cups of flour).
Area and Volume	• Linear dimensions • Square and cubic units	Linda's kitchen floor measures 12 feet by 7 feet. How many square feet of tile will Linda need to cover the whole kitchen floor? 12 feet × 7 feet = 84 square feet	The multiplicative relationship between the dimensions of two- and three-dimensional figures results in square and cubic units, respectively.

Looking Back

1. **Review the following multiplication and division problems.**
   (a) What type of problem context is represented in each problem? What is the evidence?

   (b) What are important distinguishing semantic structures of each of the problems?

Problem 1:

Each octopus has 8 legs. There were 12 octopuses at the aquarium. How many legs are there in all?

Problem 2:

Karen's garden is 6 feet by 5 feet. The area of Steph's garden is 10 times bigger than the area of Karen's garden. What is the area of Steph's garden?

Problem 3:

How many centimeters are in 140 millimeters?
10 millimeters = 1 centimeter

Problem 4:

Mrs. Cook is rearranging the books in the library. She put 7 books on the first shelf and three times as many books on the second shelf. How many books are on the second shelf?

2. **Review the student work in Figures 5.3, 5.6, 5.7, 5.9, and 5.18.**
   (a) What solution strategy is evidenced in each solution?
   (b) Where along the *OGAP Multiplication Progression* is the evidence for each piece of student work?

3. **What questions might you ask Shelby (Figure 5.18)** or how might you alter the structure of this multiplicative comparison problem to help Shelby understand the problem situation?

Instructional Link: Your Turn

Use the following questions and Table 5.1 to help think about how your instruction and math program provide students the opportunity to engage in a range of problem contexts and other problem structures (Chapter 6) for the grade level that you teach.

   1. Do you or your instructional materials provide opportunities for students to engage in the range of problem contexts targeted at your grade level?

2. Based on this analysis are there gaps between your instruction (and math program materials) and what is expected in the CCSSM and the research that supports the use of a range of problem contexts to strengthen student multiplicative reasoning?

3. If yes to question 2, what modifications should you make in your instruction?

Structures of Problems

Big Ideas

- The structure of a problem can affect the difficulty of the problem and the strategy that students use.
- Varying problem structures can increase student flexibility when solving multiplication and division problems.
- Problems can be engineered to gather specific evidence to inform instruction.

Chapter 5 investigated different problem contexts and how the different semantic structures in problems influenced student solutions. This chapter focuses on other underlying structures of multiplication and division problems that have been shown to influence the difficulty of problems, as well as the strategies students use to solve them. Problem structures refer to how problems are built. That is, how the features of the problems are organized and interact with each other. Throughout the chapter and the book there are examples of engineered problems. Engineering a problem means to alter the structures to elicit developing understandings, as well as common errors and misconceptions. As you read this chapter and think about the discussion in Chapter 5 about semantic structures, you can begin to build your understanding of what it means to engineer a problem.

There are three important ideas related to the structure of multiplication and division problems:

1. The structure of problems may play a role in the strategies students use to solve problems (OGAP, 2005).

2. The structure of problems may play a role in the level of difficulty children experience while solving problems (e.g., De Corte et al., 1988).

3. Student flexibility is increased through interaction with different problem structures and when teachers attend and respond to the difficulties they encounter while solving problems (Cobb, Yackel, & Wood, 1988; Peterson, Carpenter, & Fennama, 1989).

The influences of each of the following problem structures are discussed in this chapter:

1. The numbers in the problem
2. Number and language relationships
3. Multiplicative representations or models

The Numbers in Problems

Researchers suggest the numbers in a problem play a role in the difficulty children experience when solving problems (Bell, Swan, & Taylor, 1981; Kouba & Franklin, 1995; Steffe, 1994). This section explores the influence each of the following has on student solutions:

- The magnitude of the factors
- The presence of powers of 10 and multiples of powers of 10
- Having more than two factors in a problem

Magnitude of Factors

Magnitude refers to the size of a number. In the case of multiplication, the size of the factors are significant (e.g., 6 × 8 compared to 6 × 86), whereas in division it is the size of the divisor and dividend (e.g., 20 brownies shared by 2 students compared to 20 brownies shared by 40 students). To explore the impact of the magnitude of factors on solutions, researchers suggest giving students contextual problems where the factors are single digits and then changing the factors to multidigit numbers while the context remains the same (Greer, 1987, 1988; OGAP, 2006). Figures 6.1–6.3 provide some examples of the impact on student solutions when problems are engineered in this way.

Study Samira's solutions in Figure 6.1. As you read the problem and solutions, notice what is the same and what changed in the problem and in Samira's strategy for solving the problem.

You may have noticed that the context (e.g., equal groups) and the number of boxes of crayons (12 boxes) remained the same in both Parts A and B. What changed in Part B was the number of crayons in each box (i.e., 8 crayons in each box to 64 crayons in each box). In Part A, Samira used the distributive property to decompose 12 into 10 and 2 and multiply both by 8. From that evidence alone, one might conclude that Samira understood the problem involved multiplication and applied a strategy at the *Multiplicative* level on the progression. However, the evidence in Part B, in which the problem involved

Figure 6.1 Samira's response. Samira used the distributive property to solve the problem in Part A and inappropriately added the factors in Part B.

(a) Mark bought 12 boxes of crayons. Each box contained 8 crayons. How many crayons were there all together? Show your work.

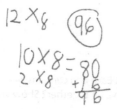

(b) John bought 12 boxes of crayons. Each box contained 64 crayons. How many crayons were there all together? Show your work.

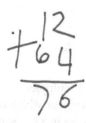

two-digit by two-digit multiplication, shows that Samira added the factors. The evidence raises questions about whether Samira understood the problem in Part A and whether in Part B her solution was highly influenced by the magnitude of the factors.

To gather additional evidence of Samira's thinking and extend her understanding, her teacher could engage her in a discussion that focuses on the context of the problem and the number relationships in the problem by asking probing questions such as:

- What is the story in Problem A and Problem B?
- How are the problems the same and how are they different?
- I noticed that you used multiplication to solve Part A and addition to solve Part B. Explain why.
- Because the problems are identical except for the number of crayons in each box, what operation would you use to solve both parts? Show me how you would solve each problem.
- How does the number of crayons in each box in Part A relate to the number of crayons in each box in Part B? How could this relationship be used to solve the problem?

Study Santiago's solutions to a similar set of problems with different size factors in Figure 6.2. In this example, unlike Samira's solution in Figure 6.1, the student recognized that both problems involved multiplication. In the first part with

Figure 6.2 Santiago's response. Santiago correctly used his understanding of the multiples of 25 to solve Part A, but incorrectly applied an algorithm to solve Part B.

(a) Samantha's class has 25 bags of cookies. Each bag contains 6 cookies. How many cookies does Samantha's class have all together? Show your work.

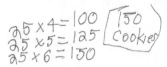

(b) Samantha's class has 25 bags of cookies. Each bag contains 24 cookies. How many cookies does Samantha's class have all together? Show your work.

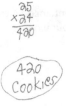

smaller numbers Santiago used his understanding of multiples of 25 to solve the problem. However, when he solved the second problem, which is identical except for the number of cookies in a bag (i.e., 6 cookies in a bag was changed to 24 cookies in a bag), Santiago unsuccessfully attempted to use an algorithm. His solution to this part indicates that he may not have developed the understanding or a strategy that would allow him to successfully solve a problem with two multidigit numbers.

Santiago's teacher could engage him in a discussion to gather additional evidence of his thinking and extend the understanding he demonstrated in Part A. One place to begin a discussion with Santiago might be with questions such as:

- Tell me about your strategy for Problem A.
- How did you know you needed to multiply to solve both problems?
- Would the strategy you used to solve Problem A work for solving Problem B?
- How might you use the strategy you used to solve Problem A to solve Problem B?
- Show me other strategies you have for solving part B.

The goal of asking Santiago these questions is to help him see how his successful strategy to solve Problem A might be built upon to solve Problem B and to see if he has effective strategies for multiplying multidigit factors.

In contrast, study Omar's solution to the same set of problems in Figure 6.3. What do you notice about his strategies for solving both problems?

You may have noticed in Figure 6.3 that Omar recognized that the number of cookies in a bag increased four times from Problem A to Problem B. He used this relationship to calculate the solution to Problem B. The engineered problem made visible that Omar recognized and used the multiplicative relationship between the factors to solve this problem. As shown in Samira's and Santiago's solutions earlier, using problems engineered like these can also make visible the struggles students are encountering when the magnitude of the numbers is changed.

Figure 6.3 Omar's response. Omar recognized the multiplicative relationship between the number of cookies in each bag in Part A and Part B and used that information to solve Part B of the problem.

(a) Samantha's class has 25 bags of cookies. Each bag contains 6 cookies. How many cookies does Samantha's class have all together? Show your work.

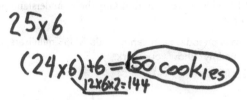

(b) Samantha's class has 25 bags of cookies. Each bag contains 24 cookies. How many cookies does Samantha's class have all together? Show your work.

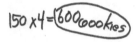

To elicit students' developing understanding and strategies, the problems in this section were engineered in two ways: 1) the magnitude of the factors was changed, but the context and language remained stable and 2) the magnitude was changed in a way that provided an opportunity for students to capitalize on the multiplicative relationships between the factors in the two problems (e.g., Omar's solution in Figure 6.3).

Influence of Powers of 10 and Multiples of Powers of 10

This section illustrates how problems can be engineered to focus on student understanding of multiplication by powers of 10 and multiples of powers of 10. Chapter 4 includes more detail about developing conceptual understanding of multiplication by powers of 10 and multiples of powers of 10. Some examples of numbers that are powers of 10 are 10, 100, 1000, etc. Some examples of multiples of powers of 10 are 50, 60, 200, 400, 3,000, etc. Study the following problem. In what way was this problem engineered to elicit student understanding of the impact of multiplying by powers of 10?

Simon knows that $40 \div 5 = 8$.
Explain how Simon can use this to find the answer to $400 \div 5$.

In this problem, you may have noticed that the dividend is 10 times as large in the second problem (400 is 40 × 10). Students should understand that this will cause the answer to the second division problem to be 10 times as large. Understanding this relationship will help students develop more efficient strategies for multiplication and division.

Study the problem in Figure 6.4. In what ways was the problem engineered to provide an opportunity for students to use their understanding of multiplication by multiples of powers of 10 and the relationships between powers of 10? Examine Taiye and Jean-Guy's responses. What is the evidence of understanding of these ideas in their work?

Figure 6.4 Taiye's and Jean-Guy's responses. Taiye shows understanding of the relationship between powers of 10 in his conversion. Jean-Guy shows understanding of multiplying by 20, a multiple of 10.

Trina bought 20 cans of soda. Each can of soda is 355 milliliters. How many liters of soda did she buy? Show your work.

1,000 milliliters = 1 liter

Taiye's response.

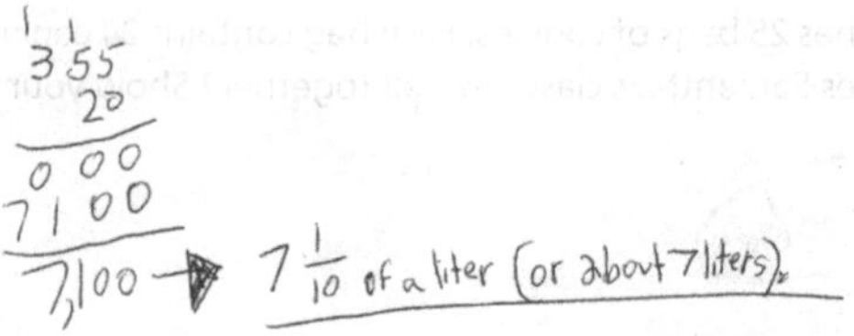

Jean-Guy's response.

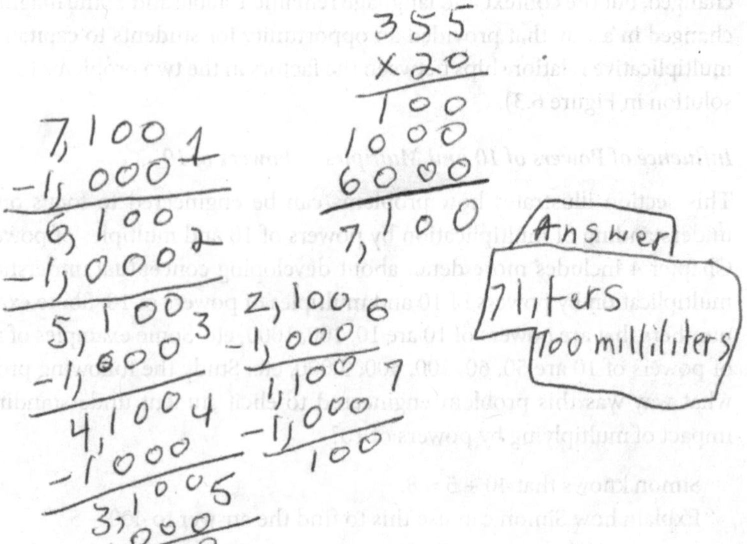

Note that this problem, unlike the previous problem about division, is cast in a contextual situation and engineered to provide the opportunity for students to recognize both multiplication by a multiple of a power of 10 (20) and the powers of 10 relationship between milliliters and liters (i.e., 1,000 milliliters ÷ 1,000 = 1 liter, or 1 liter × 1,000 = 1,000 milliliters).

The student work for this problem illustrates different evidence about each student's understanding of these concepts and relationships. Jean-Guy showed flexible and accurate multiplication by multiples of powers of 10 when applying the partial products algorithm (i.e., 20 × 5 = 100, 20 × 50 = 1,000, and 20 × 600 = 12,000). However, Jean-Guy did not use the multiplicative relationship between powers of 10 when making the conversion; rather he used repeated subtraction. In contrast, the evidence in Taiye's response shows an understanding of the 1000 to 1 relationship between 7,100 ml and 7.1 liters when she shows 7100 ml equals $7\frac{1}{10}$ liters. Ultimately we want students to recognize and apply their understanding of these multiplicative relationships when working with powers of 10 and multiples of powers of 10 as they solve problems.

Go-To See Chapter 4: The Role of Concepts and Properties for more on developing understanding of multiplication by powers of 10 and multiples of powers of 10.

The understanding of using the powers of 10 and the multiples of the powers of 10 becomes crucial for students as they begin to use the open area model, partial products, partial quotients, and the traditional US multiplication and division algorithms for solving problems with numbers that are larger than single digits. See Chapter 8 for a more detailed discussion of developing understanding of algorithms.

Three or More Factors

Students also encounter difficulty understanding problem situations when there are three or more factors in a problem, such as in the problems shown in Figure 6.5. What do you notice about Tannisha's solution?

Figure 6.5 Tannisha's response. Tannisha added 6 and 5 and multiplied that sum by 3.

Karen and Steph each have gardens.

Karen's garden is 6 feet by 5 feet. Steph's garden is three times as big as Karen's garden.
How big is Steph's garden?

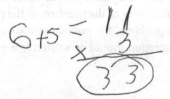

In problems with three factors students often ignore the verbal context of the problem and add the numbers (Greer, 1987, 1988). Tannisha's work reflects this research in that she added the 6 and 5, which are the dimensions of the garden. Because Tannisha then multiplied that sum by 3, we can hypothesize she did this because the problem says "three times as big." The evidence in this work would be classified on the *OGAP Multiplication Progression* as *Nonmultiplicative* because she added two of the factors. However, the student did recognize the scale factor (three times as big), and this is something that her teacher can build upon to deepen her understanding.

CCSSM: Progression of the Numbers in Problems

Table 6.1 highlights the progression of the CCSSM expectations for multiplication and division in relation to the numbers used in problems from grades 3 to 5.

Table 6.1 CCSSM expectations

	3rd Grade	4th Grade	5th Grade
Multiplication	Within 100 1-digit by multiples of 10	4-digit by 1-digit 2-digit by 2-digit	Multidigit whole numbers Powers of 10 Multiples of powers of 10 Decimals to hundredths Fractions Mixed numbers
Division	Within 100	4-digit by 1-digit	4-digit by 2-digit Powers of 10 Multiples of powers of 10 Decimals to hundredths

The Influence of Number and Language Relationships

Research suggests that the relationships of the numbers and the language in a problem can influence the operation students use on a problem and the calculation performed (e.g., Brown 1981; Vergnaud, 1983). This is not a challenge restricted to elementary students. Graeber and Tirosh (1988) found that preservice teachers wrote division equations in the order in which the numbers were presented rather than writing the equation based on the meaning of the problem. For example, when given the problem, "There are 15 friends sharing 5 pounds of cookies. How much does each friend get?" preservice teachers wrote the equation $15 \div 5$ because they thought the order of the numbers in the equation should reflect the order of the numbers in the problem. However, in this partitive division problem, the pounds of cookies are being shared by 15 friends, and so the equation should be written as $5 \div 15$.

A second challenge students face is interpreting the meaning of the words in the problem. Look at the following two problems. What do you notice about the two versions of this problem?

Problem A

Sally had 64 crayons **in each** box. She had 12 boxes of crayons. How many crayons did she have in all?

Problem B

Sally had 64 crayons **per** box. She had 12 boxes of crayons. How many crayons did she have in all?

The words in a rate problem may explicitly define the rate as a unit rate by identifying the number in each group, as in Problem A, or the words may imply the unit rate by using the word per, as in Problem B (Zweng, 1964). The expectation is that students understand "per" as a word that is used to represent a rate. This often needs to be explicitly discussed with students in multiple situations before they internalize this aspect of mathematics language. There are words like "per" that have specific mathematical meanings. There are other words like "table" and "product" that have one meaning in mathematics and a different meaning outside of mathematics, and this can also pose challenges for students.

The Influence of Multiplicative Representations and Models

Some problems are engineered with an explicit model, meaning a model is given as a part of the problem. Other problems may have an implied model, meaning a model is suggested by the context of the problem. Students should encounter problems with both implied and explicit models. It is important for teachers to be aware of how models are used or suggested in problems. Being aware of the models may help teachers make choices about what problems students will be able to solve and what strategies to expect students to use.

Figure 6.6 illustrates a problem that explicitly suggests the use of an area representation as a way to solve the problem. Ricardo used the given window as an area model, iterated that model three times, and then counted each of the individual window panes.

In contrast, the following problem implies an equal groups model because there are two groups of discrete objects being shared out into 8 equal groups:

There are 52 cards in one deck of cards.

Two decks of cards are combined to play a game.
Eight people are playing the game.

How many cards will each person receive if all the cards are passed out equally?

Figure 6.6 Ricardo's response. Ricardo used the area model to iterate the given window and accurately determined the number of panes in three windows.

The Smiths have 3 windows on the front side of their house like the picture below. Each window contains many panes of glass.

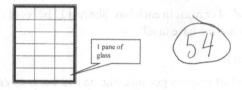

How many panes of glass are there in the 3 windows on the front of the Smiths house?

Show your work.

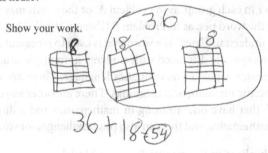

The language of the problem does not directly indicate which representation to use when solving the problem. However, the context may influence the visual model that students imagine or draw to solve the problem. Study the work of Jenna, Mohammed, and Luciana (Figures 6.7–6.9). What do you notice about each of their solutions?

You may have noticed Jenna (Figure 6.7) drew 8 groups and then shared out the cards by ones consistent with the contextual situation. Jenna sketched each card that was shared out to the 8 children. Mohammed (Figure 6.8) did not need to draw a visual model and used multiplication and division to express his solution to the problem. Luciana (Figure 6.9) began by recognizing that two decks of cards is 104 cards. She then made 8 groups, distributed 10 to each group, and then the remaining 24 in equal groups of 3.

If you use this problem in your classroom, you might select work similar to Jenna, Mohammed, and Luciana to begin a discussion about the problem. To move student understanding forward, you could ask questions of the whole class such as:

- How are these strategies similar to and/or different from each other?
- Jenna and Luciana both made equal groups. Where can you see Luciana's groups in Jenna's drawing?

- How did Luciana represent the equal groups?
- What connections do you see between Luciana's solution and Mohammed's equations?

One goal of this discussion is to expose students to ways the implied model can be used to solve the problem. A second goal is to help students see connections between strategies that are at different levels on the *OGAP Multiplication Progression*.

Figure 6.7 Jenna's response. Jenna drew 8 groups and shared out the cards equally.

Figure 6.8 Mohammed's response. Mohammed used multiplication and division to solve the problem.

$$52 \times 2 = 104$$

$$104 \div 8 = 13$$

Figure 6.9 Luciana's response. Luciana shared out cards in equal groups by representing the cards with numbers rather than drawing each card.

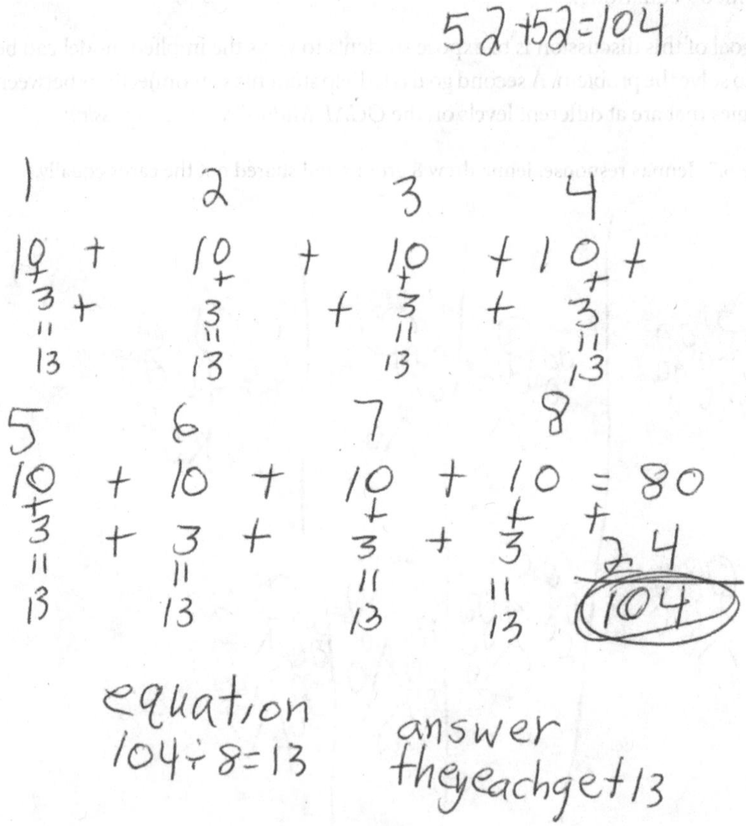

$$52 + 52 = 104$$

$$
\begin{array}{cccc}
1 & 2 & 3 & 4 \\
10\ + & 10\ + & 10\ + & 10\ + \\
3\ + & 3 & +\ 3\ + & 3 \\
11 & 11 & 11 & 11 \\
13 & 13 & 13 & 13 \\
\\
5 & 6 & 7 & 8 \\
10\ + & 10\ + & 10\ + & 10\ =\ 80 \\
3\ + & 3\ + & 3\ + & 3\ + \\
11 & 11 & 11 & 24 \\
13 & 13 & 13 & \boxed{104} \\
\end{array}
$$

equation
$104 \div 8 = 13$

answer
theyeachget13

Engineering Problems

As stated in the beginning of the chapter, engineering a problem means to alter the structures to specifically elicit developing understandings, as well as common errors and misconceptions, based upon findings in math education research. Every problem in this chapter (and throughout the book) is an example of an engineered problem.

Teachers who have participated in OGAP studies consistently say that having knowledge of the math education research about problem structures and progressions has provided a new lens through which to make instructional decisions. Understanding how problems are engineered can inform instruction in a number of ways, including helping to:

1. Anticipate the kinds of solutions students might generate and the challenges they might experience
2. Consider and plan responses to students' solutions

3. Select problems based on the goals of the lesson
4. Balance the types of problems and the structures in the problem that students encounter
5. Analyze instructional materials to engage students in a variety of problem structures
6. Analyze the next textbook lesson to address concepts students are struggling with

Engineering problems by itself will not do all the work of strengthening students' understanding, but carefully selected problems in combination with classroom discussions can help develop students' conceptual understanding and multiplicative reasoning.

Summary

- The structure of a problem can be engineered to elicit student understanding of foundational concepts, as well as common errors and misconceptions.
- Structures of problems include the complexity of the numbers, the number and language relationships, and the implied or explicit models in a problem.
- The information about student understandings gathered from engineered problems can be used to guide instruction.

Looking Back

1. **Explicit and Implied Models**: We learned in this chapter that multiplication and division problems can have a particular associated visual model. These models can be either explicit or implied. Examine the five problems shown in Figure 6.10 and:

   (a) Identify the model associated with each problem

   (b) Explain whether the model is explicit or implied

Figure 6.10 Five problems.

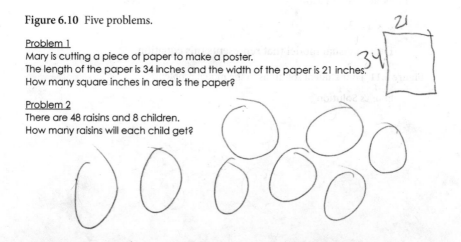

Problem 1
Mary is cutting a piece of paper to make a poster.
The length of the paper is 34 inches and the width of the paper is 21 inches.
How many square inches in area is the paper?

Problem 2
There are 48 raisins and 8 children.
How many raisins will each child get?

Problem 3

Here is a diagram of Linda's floor.

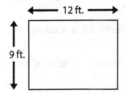

How many one-foot square tiles are needed
to cover Linda's bedroom floor?

Problem 4

Sam found the total number of stars by counting each one.
Sam counted 16 stars.

Problem 5

Show Sam another way to find the total number of stars.

2. **Analyzing Student Solutions:** The following problem is an example of
   a problem that is not associated with any particular model. This type of
   problem can help teachers understand how students are visually interpreting multiplication situations. Examine the problem and study each student
   solution (Figure 6.11). For each solution identify:
   (a) The type of model the student drew
   (b) The student's developing understanding(s) suggested by the solution
   (c) Any underlying issues or concerns to consider in future instruction

   Model Problem

   Look at this equation:

   $8 \times 4 = 32$

   Draw a visual model that represents this equation.

Figure 6.11 Four student solutions to the model problem.

Imani's Solution

Jane's Solution

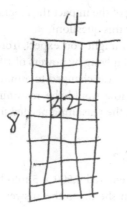

Joaquin's Solution

Trina's Solution

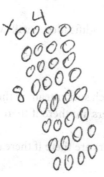

3. **Using Understanding of Multiplying and Dividing by Multiples of 10 to Solve Related Problems:** Examine the problem and answer the questions that follow.

The Calculator Problem

Matilda accidentally entered 4,500 ÷ 5 into her calculator instead of 4,500 ÷ 10. What can she do to the answer of 4,500 ÷ 5 to find the answer to 4,500 ÷ 10?

(a) How might a student who has an understanding of place value, as well as an understanding of the impact the magnitude of a divisor has on the quotient, answer this question?
(b) What specific errors might you expect from students who are still making sense of multiplying by powers of 10?
(c) Imagine the question was changed to a multiplication situation like the next example. How would the reasoning needed to solve this problem differ from the reasoning needed to solve the original division question?

The Calculator Problem 2

Matilda accidentally entered 4,500 × 5 into her calculator instead of 4,500 × 10. What can she do to the answer of 4,500 × 5 to find the answer to 4,500 ×10?

4. **Try This with Your Class:** Administer a question like the following soccer problem to help you collect information on how the magnitude of numbers in a multiplication problem might affect student solutions. Use the following prompts to guide your analysis of your students' work:
   - Which students answered both parts of the question correctly using the same strategy? Were these correct strategies *Additive, Early Transitional, Transitional,* or *Multiplicative*? What is the evidence?
   - Which students used a less efficient strategy on Part B than on Part A?
   - Which students used a *Nonmultiplicative* strategy on Part B but not on Part A? Did any students use *Nonmultiplicative* strategies on both parts of the question?
   - What instructional modifications will you make based on this information?

The Soccer Problem

There are 16 players on each soccer team in the Smithville Soccer League.
(a) How many total players are there if there are 8 teams in the league? Show your work.
(b) How many total players are there if there are 24 teams in the league? Show your work.

5. Review the student work in Figures 6.1–6.9.
   (a) What strategy is evidenced in each solution?
   (b) Where along the *OGAP Multiplication Progression* is the evidence for each piece of student work?

Instructional Link: Your Turn

Use the following questions to help you examine ways in which your math instruction and math program provide students opportunities to work with problems containing various structures.

1. What are the ways that you and your instructional materials provide opportunities for your students to engage with problems that vary in relation to the:
   - Magnitude of numbers?
   - Number and language relationships?
   - Types of models, implied or explicitly described?
2. Given your findings in Question 1, identify instructional modifications that you can make to ensure that your students have regular opportunities to work with a variety of multiplicative problems that differ in structure.

Developing Whole Number Division

Big Ideas

- Students should solve a range of division problems involving different problem structures, different contexts, and different contextual situations.
- A focus on the meaning of the quantities in division problems is critical for understanding the contextual situation and the interpretation of remainders (Fosnot & Dolk, 2001).
- The *OGAP Division Progression* can be used to gather evidence of students' developing understanding of division and help facilitate instructional decision-making.

Students' conceptions and competencies develop over long periods of time, through experience with a large number of situations, both in and out of school. When faced with new situations (new domain, new relationship, new numerical data), they use the knowledge that has been shaped by their experience with simpler and more familiar situations and try to adapt it to this situation (Vergnaud, 1988, p. 141).

Development of understanding and fluency with whole number division is not something that begins with formal instruction in second and third grade. Rather, as Vergnaud (1988) suggests, "whole number division understanding and competencies, like other new concepts, are developed over time and are built on earlier school and non-school experiences" (p. 141).

The foundations of understanding of division as a concept come from equal sharing experiences during preschool years. For example, children share with their peers by partitioning food portions into equal parts or dealing out equal shares of discrete objects before they even enter formal schooling situations. More formal instruction at grade 3 builds on these sharing experiences as students solve division problems such as, "Three children share 6 brownies equally. How many brownies does each student get?"

Division as an operation builds from students' developing understanding of multiplication and the inverse relationship between multiplication and division. Students begin to understand that division is related to multiplication in much the same ways that subtraction is related to addition.

They understand, for example, that 6 brownies shared among 3 people results in 2 brownies per person. They see that there are two related equations: one division and the other multiplication.

6 brownies ÷ 3 people = 2 brownies per person
3 people × 2 brownies per person = 6 brownies

 Go To For more about the inverse relationship between multiplication and division go to Chapter 4: The Role of Concepts and Properties.

As soon as formal instruction begins on division, *partitive* and *quotative* situations should be introduced and taught simultaneously. These terms will be fully exemplified and defined in the next section of this chapter. Over time students should be introduced to, and solve, division problems in a range of contexts and contextual situations with larger dividends and divisors, using understanding of place value and an understanding of the inverse relationship between multiplication and division.

This chapter focuses on understanding partitive and quotative division, interpreting remainders, and the different levels on the *OGAP Division Progression.*

Partitive and Quotative Division

Researchers indicate that students should solve both partitive and quotative division problems, as well as those division problems involving area, multiplicative comparisons, scale, volume, and measurement conversion problems. However, some researchers have found that there is a greater instructional focus on partitive problems than quotative problems (Greer, 1992). To begin to understand the difference between partitive and quotative division problems, solve Problems 1 and 2. How are these problems alike and how are they different?

Problem 1

Twenty students equally share 100 jellybeans. How many jellybeans does each student get?

Problem 2

There are 100 jellybeans. Each package holds 20 jellybeans. How many packages of jellybeans are there?

You probably noticed that both problems are arithmetically equivalent ($100 \div 20 = 5$). However, these are very different problem situations involving different semantic structures.

Go To For a discussion explaining semantic structures and their impact of solving multiplication and division problems, go to Chapter 5: Problem Contexts.

Problem 1 is a sharing or partitive division problem. That is, the problem is asking how many jellybeans are in each group. We know the total number of jellybeans (100) and we know the number of groups (20 students).

100 jelly beans ÷ 20 students = 5 jelly beans per student
Total number of objects ÷ number of groups = number in each group

Problem 2 is a quotative division problem. In this case we are given the total number of jellybeans (100) and the number of jellybeans in each package (20 jellybeans per package). The problem is asking for the number of packages (groups) of jellybeans with 20 jellybeans in each package.

100 jelly beans ÷ 20 jelly beans per package = 5 packages
Total number ÷ number in each group = number of groups

The difference in the semantic structures in quotative and partitive division problems often affects the strategies students use to solve each type of division problem as they are developing understanding of division. Also, the quantities in the solutions to partitive and quotative problems have different meanings. Partitive division problems result in the number in each group (the composite unit). Quotative division problems result in the number of groups.

Beginning Strategies Evidenced in Partitive and Quotative Division Problems

Figure 7.1 contains two responses with evidence of dealing or sharing out—often the first strategy students use to solve partitive division problems. Dealing or sharing out is evidence that these students understood the concept of dividing a quantity into four equal groups as pictured.

In quotative division problems the number in each group (the composite unit) is known, but not the number of groups. Students cannot share out into groups if they don't know how many groups to make. Instead, students tend to use additive strategies to either add up the number in each group until they arrive at the total or subtract from the total until they arrive at zero. Examples of these two strategies are shown in Figure 7.2.

The solutions in Figures 7.1 and 7.2 are classified as *Additive Strategies* on the *OGAP Division Progression* because students are counting, adding, or

Figure 7.1 Two responses to a partitive division problem in which the responses contain evidence of dealing or sharing out. Response A shows evidence of sharing out by ones, whereas Response B shows evidence of sharing out by groups of tens and then threes.

The fourth grade class earned $132 from a bake sale. The class decided to donate an equal amount of the money to four different organizations. How much money did the class donate to each organization? Show your work.

Response A—sharing out by ones   **Response B—sharing out by groups of tens and then groups of 3S**

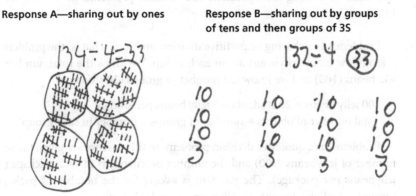

Figure 7.2 Two correct solutions to a quotative division problem in which the responses contain evidence of repeated addition (A) and repeated subtraction (B).

A class of 234 students is going on a field trip. The buses used to transport the students each seats 28 students. How many buses will be needed to take all the students on the field trip? Show your work.

Response A—repeated addition

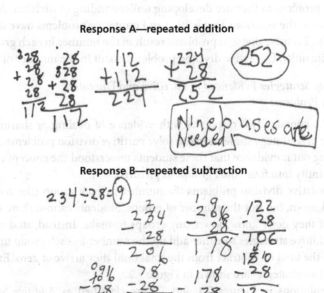

Response B—repeated subtraction

subtracting the amount in each group. As mentioned, it is the structure of these division problems that influences these beginning strategies. As students begin to solve partitive division problems they tend to share out because they are given the number of groups and the total number of objects that need to be shared. In contrast, in quotative division problems they are given the total number of objects and the number in each group, so as students begin to solve quotative division problems they tend to either repeatedly subtract the number in each group or build up to the total. It is important for students to engage in both partitive and quotative division, as well as engage in division problems that are neither partitive nor quotative that involve area, volume, conversions, and multiplicative comparisons.

Interpreting Remainders

Studies have shown that students often select the correct operation (division) when solving division word problem involving remainders but do not return to the problem situation or context to interpret the remainders appropriately (Silver, Shapiro, & Deutsch, 1993). This finding has important instructional implications as students engage in different division contexts and contextual situations. Samantha's response in Figure 7.3 exemplifies this point. Samantha's use of the open area model and calculations are correct, but the lack of labels indicating that each person received 9 pens with 10 pens left over provides evidence that she may not have returned to the problem context after obtaining her answer.

Figure 7.3 Samantha's response. Samantha accurately calculated an answer of 9 R 10. However, the lack of labels in her answer may be evidence that she did not return to the problem context to make a decision about how to treat the remainder.

There are 145 pens to share equally with 15 people. How many pens will each person get?

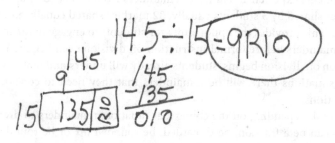

In contrast, the solutions in Figure 7.2 show evidence of returning to the problem situation to determine what to do with the remainder. In both cases the students reasoned that an extra bus was needed to transport the remaining students. Figure 7.4 is a schematic that shows the idealized path to a successful solution to a division problem with a remainder.

Figure 7.4 Schematic showing the idealized path to a division problem that involves remainders. (Adapted from Silver et al., 1993, p. 120.)

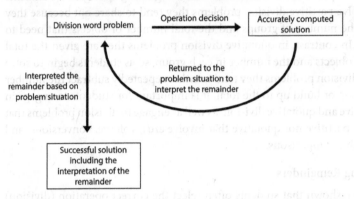

It is the contextual situations (e.g., sharing 6 balloons compared to sharing brownies) of a division problem that determines how remainders should be interpreted. "Students should not just think about remainders as 'R3' or '3 left over'. Addressing what to do with remainders should be central to the teaching of division" (Van de Walle, Karp, & Bay-Williams, 2012, p. 161). Additionally, "children should encounter division problems involving remainders from the time they begin work on division ideas" (Reys, Lindquist, Lambdin, & Smith, 2009, p. 256).

These two ideas have implications for instruction with regard to teaching and learning division and the meaning of remainders. That is, students should encounter division with remainder problems as they learn about division and be asked to think about what to do with that remainder. One might question whether it is too early to deal with remainders when students are first learning about division. However, think about the many real-life situations students experience that have remainders from a very early age (e.g., 12 cookies shared by 5 students equally; 32 marbles shared equally among 5 children). This provides the opportunity for students to engage in the meaning of remainders before formal instruction on division even begins. When instruction on division begins, students already will understand that in some division situations there will be remainders that they need to consider in their solutions.

In general, depending on the contextual situation, remainders to division problems can be a fraction, be discarded, be rounded up, or be rounded to the nearest whole number. Table 7.1 provides some problem examples with explanations on how the contextual situation drives the decision about how to treat the remainder in division problems. Read and solve each problem. Then think about how the contextual situation influenced how the remainder was interpreted.

Table 7.1 Problem examples of how the contextual situation drives the decisions about how to treat the remainder in the solution.

Problem	Contextual Situation that Influences the Interpretation of the Remainder	Solution
Remainder is divided into a fractional part		
Thirty-six brownies are shared equally with 8 students. How many brownies does each student get? (Partitive division)	This is a partitive division problem involving equal groups. Because the object being shared can be partitioned into fractional parts. 36 brownies ÷ 8 students = x brownies per student	4½ brownies per student
A walking race is 15 miles long. If I walk 4 miles per hour, how many hours will it take me to walk the 15 miles? (Quotative division)	This is a quotative division problem involving a rate, distance, and time. The problem asks for the number of hours it will take to complete the race. Because hours are continuous measures, the remainder can be expressed as a fraction of an hour. 15 miles ÷ 4 miles per hour = x hours	3¾ hours
Remainder is discarded		
Thirty-six balloons are shared equally among 8 students. How many balloons does each student get? (Partitive division)	This is a partitive division problem involving equal groups. Because the objects being shared (balloons) cannot be partitioned, the remaining balloons are discarded or given away. 36 balloons ÷ 8 students = x balloons per student	4 balloons per student; 4 balloons discarded
A decoration needs 15 inches of ribbon. I have 326 inches of ribbon. How many decorations can I make? (Quotative division)	This is a quotative division problem involving equal measures. Because the length of ribbon needed to make one decoration is a set amount, any ribbon not used is discarded. 326 inches of ribbon ÷ 15 inches per decoration = x decorations	21 decorations with 11 inches of ribbon not used
Solution rounded up		
A class of 234 students is going on a field trip. The buses used to transport the students each can hold 28 students. How many buses will be needed to take all the students on the field trip? (Quotative division)	This is a quotative division problem involving equal groups. Because all the students are going on the trip, the answer is rounded up to account for all the students. 234 students ÷ 28 students per bus = x buses	9 buses
Rounded to the nearest whole number		
In 2005 Vermont received 8 inches of snow in a snowstorm. On the same date in 1978 Vermont received 34 inches of snow. About how many times more inches of snow did Vermont receive in 1978 than on the same date in 2005?	This is a multiplicative comparison problem asking for an estimate of the scale factor. Therefore, the solution should be rounded to the nearest whole number. 34 inches ÷ 8 inches = x times the amount of snow	About 4 times as much snow in the 1978 blizzard than the 2005 storm.

Notice that it is the contextual situation of the problem that influences the decision about how to interpret the remainder. There are no set rules to impart to students about how to interpret remainders other than focusing on the situation in the problem and using that to guide decisions about how to treat the remainder. One way to help students think more deeply about the contextual situation is to use the *Word Problem Strategy* described in Chapter 5.

Another way to help students interpret remainders is by having them solve application problems in which they must select between different interpretations of the remainders. See the problems in Figures 7.5 and 7.6. This type of problem is engineered to force students to think about the meaning of the remainder and can provide a platform for meaningful discussion about interpreting remainders. The student solutions in Figures 7.5 and 7.6 are different interpretations of the remainders in the same problem that could be used for discussion. Review Caleb's and Omar's responses in Figures 7.5 and 7.6 and think about how you might engage students in a discussion about the remainders in this problem.

Figure 7.5 Caleb's response. Caleb determined that each student would get 2 brownies with 10 brownies left over without considering that brownies can be divided into fractional pieces.

Twenty-five students are sharing 60 brownies equally. How many brownies should each student get?

25 Brownes 2×25 is 50 Brownies
2 Children can't eat ten Brownies
Each student gets 2
Remainder of 10 Brownies

B. **For each statement** circle correct or incorrect to indicate how many brownies each student should get.

E) Two brownies each	Correct	Incorrect
F) Two and two-fifths of a brownie each	Correct	Incorrect
G) 15 students get 2 brownies each, and 10 students get 3 brownies each	Correct	Incorrect

Figure 7.6 Omar's response. Omar correctly determined that each student would get 2 2/5 brownies

Twenty-five students are sharing 60 brownies equally. How many brownies should each student get?

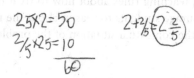

B. **For each statement** circle correct or incorrect to indicate how many brownies each student should get.

E) Two brownies each	Correct	Incorrect
F) Two and two-fifths of a brownie each	Correct	Incorrect
G) 15 students get 2 brownies each, and 10 students get 3 brownies each	Correct	Incorrect

A strategy to use these solutions to guide a discussion about the interpretation of remainders is to project both Caleb's and Omar's solutions in front of the class and ask a series of guiding questions that starts with the context of the problem and extends to the bigger idea of interpreting remainders. Some sample guiding questions follow.

Sample guiding questions:

1. Look at both Omar's and Caleb's solutions. How are they alike and how are they different?
2. Under what conditions would Omar's solution be correct?
3. Under what conditions would Caleb's solution be correct?
4. Neither Omar nor Caleb thought that answer (G) was correct. Under what conditions would (G) be a correct response?
5. In this context the extra brownies can be divided into fractional pieces. What are other situations that involve sharing that the extras can be divided into fractional pieces? Are there times when dividing up the extras would not make sense?
6. What are sharing situations where the leftovers cannot be divided up? For each of the situations what are ways that the leftovers can be treated?

These examples point to two important instructional ideas with regard to helping students interpret remainders: 1) students should pay careful attention to the meaning of remainders in a range of problem situations and contexts, and not just focus on the development of the procedure to divide and 2) teachers should avoid teaching rules about how to treat remainders. Rather, teachers should encourage students to reason about the meaning of the remainder in relation to the contextual situation of the problem and the quantities involved.

The *OGAP Division Progression*

The same principles that apply to the *OGAP Multiplication Progression* discussed in Chapter 3 apply to the *OGAP Division Progression*.

- The evidence in student work falls along the levels from *Nonmultiplicative to Multiplicative* strategies.
- The progression is not linear. Rather, strategies students use move up and down on the progression depending on the problem situations, contexts, and the problem structures.
- Students will be at different places at different times.
- The progression provides instructional guidance for moving student strategies along the continuum.
- The progression is not evaluative.
- Collection of *Underlying Issues and Errors* is important (such as interpretation of remainders).

The progression levels represent the continuum of evidence from *Nonmultiplicative* to *Multiplicative Strategies* that is visible in student work as students develop their understanding and fluency for whole number division. The levels are at a grain size that is usable by teachers to gather actionable evidence to influence instructional decision-making.

Open to pages 2 and 3 of the *OGAP Multiplicative Reasoning Framework*. You will notice that the *OGAP Multiplication Progression* is on page 2 and the *OGAP Division Progression* is on page 3. Notice that the levels along the division progression are the same as they are with multiplication: *Multiplicative Strategies; Transitional Strategies; Early Transitional Strategies; Additive Strategies; Early Additive Strategies; Nonmultiplicative Strategies*. In addition, when using the *OGAP Division Progression*, it is important to consider the levels on the progression, as many students will use multiplicative strategies to solve division problems. For example, in Figure 7.6 Omar used multiplication to determine that each student would receive $2\frac{2}{5}$ of a brownie.

The design of the *OGAP Division Progression* is to provide a learning path based on mathematics education research that develops procedural fluency with understanding. The *Transitional Level* in both progressions is the bridge between additive reasoning and strategies that reflect procedural fluency with understanding at the multiplicative level.

From Early Additive *to* Additive *Strategies*

As discussed earlier in the chapter, when students first engage in division problems, they often draw on additive strategies to solve them. These strategies include sharing out by ones or in groups as shown in Figure 7.1 or use of repeated subtraction as shown in Figure 7.2.

Early Additive strategies involve sharing out and counting by ones. As students' understanding of numbers develops, they are able to conceptualize and work with units larger than one, for example, sharing out and counting by larger groups (Figure 7.1) or repeated addition or subtraction (Figure 7.2). These strategies that involve unitizing with units larger than one are considered *Additive*.

From Additive *to* Transitional

The *Additive* solutions of repeated addition and repeated subtraction evidenced in Figure 7.2 are accurate, including the interpretation of the remainder. However, they are not efficient, nor do they show evidence of understanding the multiplicative relationships so important to being procedurally fluent with understanding. The focus at the *Transitional* level is to bridge students from additive reasoning to procedural fluency with understanding at the multiplicative level.

At the *Early Transitional* level building-up strategies and skip counting are evidenced as students move away from their counting strategies to operate on groups. Ashley's response in Figure 7.7 combines skip counting by

Figure 7.7 *Early Transitional* strategy. Ashley's response shows evidence of using skip counting to accurately determine the amount donated to each group.

> The fourth grade class earned $132 from a bake sale. The class decided to donate an equal amount of the money to four different organizations. How much money did the class donate to each organization? Show your work.

$$132 \div 4 = 33$$

4, 8, 12, 16, 20, 24, 28, 32, 36, 40, 44, 44,
48, 48, 52, 56, 60, 64, 68, 72, 76, 80, 84,
88, 92, 96, 100, 104, 108, 112, 116, 120,
124, 128, 132

fours and then counting the number of skips (33) to determine the amount to donate to each group. Ashley's solution is correct. However, there are no units labeled, so one is not certain that she understands that her answer, 33, refers to dollars.

Important in the move from *Early Transitional* to *Transitional Strategies* is understanding and fluency with the use of the inverse relationship between multiplication and division. For example, in Eli's solution in Figure 7.8 there is evidence that he understands the inverse relationship between multiplication and division when he used multiplication to solve this division problem. However, the solution is at the *Transitional Strategy* level because it was not efficient.

Figure 7.8 *Transitional Strategy*. Eli's solution shows evidence of understanding the inverse relationship between multiplication and division by using multiplication to solve a division problem. It is classified as *Transitional* because the solution was not efficient.

The fourth grade class earned $132 from a bake sale. The class decided to donate an equal amount of the money to four different organizations. How much money did the class donate to each organization? Show your work.

At the *Transitional* level there may also be evidence of using the partial quotient algorithm, but not efficiently. Roberta's solution in Figure 7.9 shows use of partial quotients, but the continuous use of 3 as a partial quotient without considering more efficient dividends makes the strategy inefficient.

 See Chapter 8 for more on making sense of the partial quotient algorithm.

Figure 7.9 *Transitional Strategy*. Roberta's response shows evidence of using the partial quotient algorithm. Note that Roberta made sense of the remainder in this problem situation.

> There are 15 teams in a town soccer league. The league made sure that about the same number of players are on each team. There are 200 players in the league. How many players will be on each team?

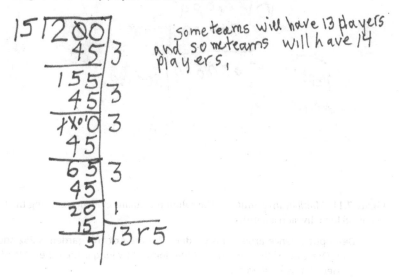

Multiplicative Strategies

At this level, students use efficient strategies (e.g., partial quotients, traditional US algorithm), as well as appropriately understanding place value and the inverse relationship between multiplication and division, to solve division problems. By sixth grade students should be using *Multiplicative Strategies* to solve whole number division problems involving equal groups, equal measures, measurement conversions, multiplicative comparisons, unit rates, factors and multiples, rectangular area, and volume problems.

Solutions at the *Multiplicative Strategy* level include efficient partial quotients (Figure 7.10), the traditional US long division algorithm (Figure 7.11), and efficient use of the inverse relationship between multiplication and division (Figure 7.12).

Go To Chapter 8: Understanding Algorithms provides an in-depth discussion about how area models can be used to make sense of the traditional US division algorithm.

Figure 7.10 *Multiplicative Strategy.* The solution contains efficient use of partial quotients.

In 2004 there were about 5,760,000 motorcycles in the United States. That is about 10 times more than the number of motorcycles in 1960. About how many motorcycles were there in the United States in 1960?

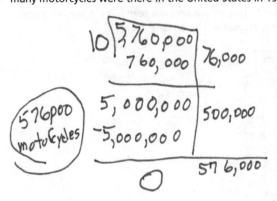

Figure 7.11 *Multiplicative Strategy.* The solution contains evidence of using the traditional US long division algorithm.

Dean put a fence around his garden. The area of the garden is 252 square feet. One side of the garden is 18 feet long. How long is the other side of the garden? Show your work.

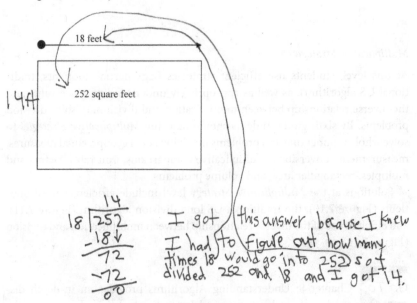

Figure 7.12 The solution contains evidence of efficiently using the inverse relationship between multiplication and division.

> The average price of a home in 1960 was $12,700. The average price of a home in 2008 was $238,880. About how many times more was the cost of the home in 2008 than in 1960?

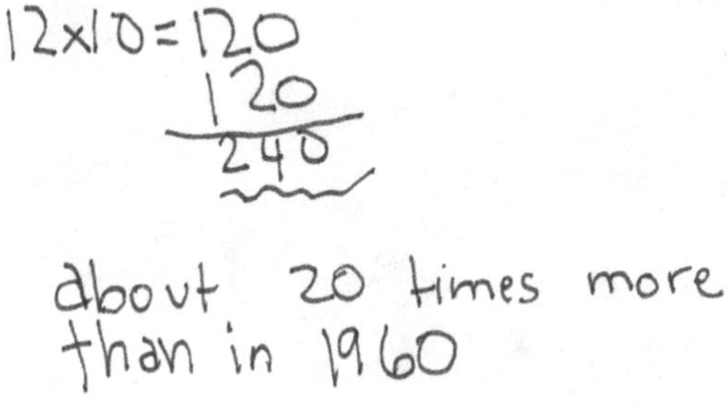

$12 \times 10 = 120$

$$\begin{array}{r} 120 \\ \underline{120} \\ 240 \end{array}$$

about 20 times more than in 1960

Nonmultiplicative Strategies

At the bottom of the progression notice a box labeled *Nonmultiplicative Strategies*. As students engage in new topics or are just beginning division concepts they often add or multiply the dividend and the divisor, use the incorrect operation, or guess. Figures 7.13 and 7.14 are some examples of solutions with evidence of *Nonmultiplicative Strategies*.

Figure 7.13 *Nonmultiplicative Strategy*. The solution contains evidence of multiplying the dividend and the divisor instead of determining the scale factor between 19 inches and 194 inches.

> A piece of elastic that is 19 inches long can be stretched to 194 inches. How many times its original length can it be stretched? Show your work.

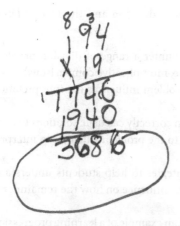

Figure 7.14 *Nonmultiplicative Strategy.* The solution contains evidence of subtracting the divisor from the dividend instead of determining the scale factor between the average cost of a home in 1960 and 2008.

> The average price of a home in 1960 was $12,700. The average price of a home in 2008 was $238,880. About how many times more was the cost of the home in 2008 than in 1960?

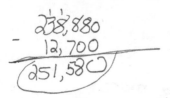

Go To See Chapter 2: The OGAP Multiplication Progression for more on how the sorting and analysis of evidence process is used to inform instruction for both the *OGAP Division Progression* and *OGAP Multiplication Progression.* As one collects evidence of underlying issues for division, particular attention should be paid to the interpretation of the remainder and units.

The progression, as you can see, shows that student strategies move from counting by ones, to counting by groups, to operating on groups multiplicatively, to using models, to developing efficient strategies that are built on conceptual understanding. Although students first engage in division using additive strategies, the *Transitional* level is the bridge from these additive strategies to the efficient strategies at the *Multiplicative* level.

Chapter Summary

This chapter focused on whole number division and the *OGAP Division Progression.*

- Across time students should encounter a range of division problem situations (e.g., partitive, quotative, multiplicative comparisons).
- The contextual situation of the problem influences the interpretation of remainders.
- Research shows that students often correctly calculate solutions to division problems, but do not return to the problem situation to interpret the meaning of the remainders.
- Instruction should focus on strategies to help students understand the context of the problem and its influence on how the remainder is interpreted.
- The *OGAP Division Progression* is an example of a learning progression founded on mathematics education research and written at a grain size

that is usable by teachers and students in a classroom across a range of multiplicative reasoning concepts.

• The *OGAP Division Progression* was specifically designed to inform instruction and monitor student learning from a formative assessment perspective.

Looking Back

1. **Partitive and Quotative Division Problems**: It is important that students have ample opportunities to solve both partitive and quotative multiplication problems.

   (a) For each of the following four problems:
   - Indicate whether the problem is a partitive or quotative division problem
   - Indicate the quantities given in the problem and the unknown quantity

       Problem 1: Ann and Billy baked 60 cookies for the bake sale. They put an equal number of cookies on 5 plates. How many cookies did they put on each plate?

       Problem 2: The class collected 65 pounds of food for their food drive. The students filled each box with 5 pounds of food. How many boxes did they fill?

       Problem 3: Jillian has 45 ounces of apple juice. The apple juice is packaged in bottles that each contains 15 ounces. How many bottles of apple juice does Jillian have?

       Problem 4: Three children equally share 45 ounces of apple juice. How much juice does each child get?

   (b) Two distinct division problems can usually be created from one multiplication problem. All three of these problems utilize the same context and the same quantities. Write two different division problems related to the following multiplication problem. Identify each division problem you wrote as either partitive or quotative.

The Jellybean Problem

Five friends buy enough jellybeans so that each friend gets 6 ounces of jellybeans. How many ounces of jellybeans do the 5 friends buy?

2. **Student solutions to partitive and quotative division problems:**

   (a) The bake sale problem in Figure 7.1 and also shown here is an example of a partitive division problem. Use the *OGAP Division Progression* to help you analyze Responses A and B in Figure 7.1. What level of the progression best describes each strategy? What is the evidence that supports this analysis?

The Bake Sale Problem

The fourth grade class earned $132 from a bake sale.
They donated an equal amount to four different organizations.
How much money did the class donate to each organization?

3. Two student solutions to a quotative division problem are shown here.
Use the *OGAP Division Progression* to help you analyze Joy's and Andrew's
solutions (Figures 7.15 and 7.16) to the music festival problem. What
level of the progression best describes each strategy? What is the evidence
that supports this analysis?

The Music Festival Problem

Buses will take 2881 participants from the parking lot to the music
festival site.
    One bus holds 67 participants.
    How many bus trips will it take to get all participants from the parking
lot to the music festival site?

Figure 7.15 Joy's solution to a quotative division problem.

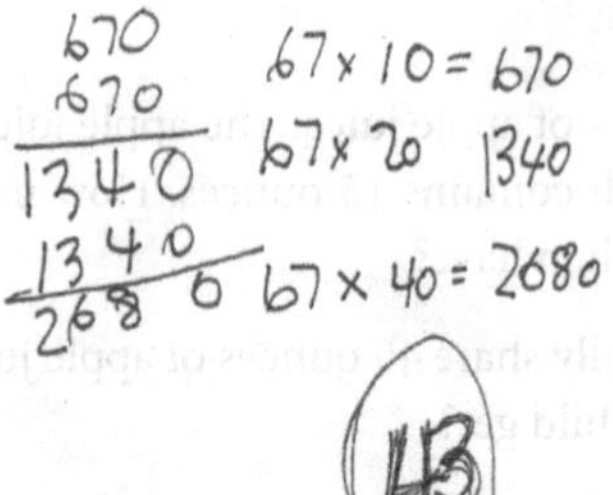

Figure 7.16 Andrew's solution to a quotative division problem.

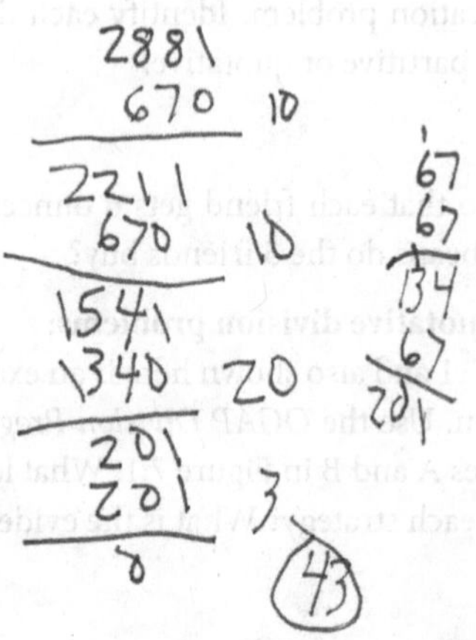

(c) What are some differences between Responses A and B to the partitive problem in Figure 7.1 and Joy's and Andrew's solutions to the quotative problem?

4. **Division Problem Involving Multiplicative Comparison**: Multiplicative comparison is an example of a context that is neither partitive nor quotative. Study the multiplicative comparison problem next and the five student solutions that follow (Figures 7.17–7.21).

(a) Use these solutions and the *OGAP Multiplication Progression* to analyze the evidence in each of the solutions. Record your analysis on a copy of the *OGAP Evidence Collection Sheet* shown in Figure 2.24. For each solution identify:

- The level on the progression the evidence in the solution is located. What is the evidence?
- Any *Underlying Issues or Errors*.
- Whether or not the answer is correct. Highlight the student solutions that are incorrect.

Multiplicative Comparison Division Problem

(b) Based on your analysis, identify understandings that can be built upon and potential strategies you could use to help each student build understanding toward the next level on the *OGAP Division Progression*.

In 2004 there were about 5,760,000 motorcycles in the United States. This is about 10 times more than the number of motorcycles in 1960. About how many motorcycles were there in the United States in 1960?

Figure 7.17 Jon's response to a multiplicative comparison division problem.

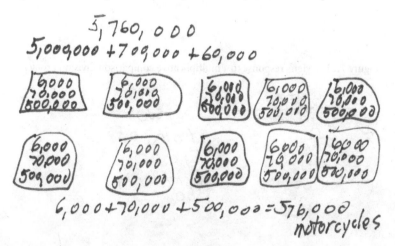

Figure 7.18 Emma's response to a multiplicative comparison division problem.

5,076,000
10 ⟌ 5,760,000
50
076
70
060
60
00

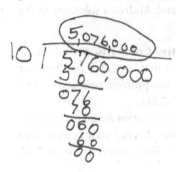

Figure 7.19 Kat's response to a multiplicative comparison division problem.

? x 10 = 5760000
576000
576000
x 10
5760000

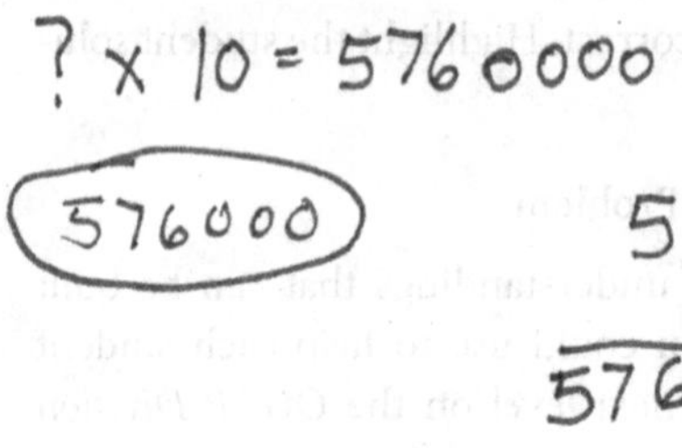

Figure 7.20 Ben's response to a multiplicative comparison division problem.

5,760,000
− 10
5,759,990

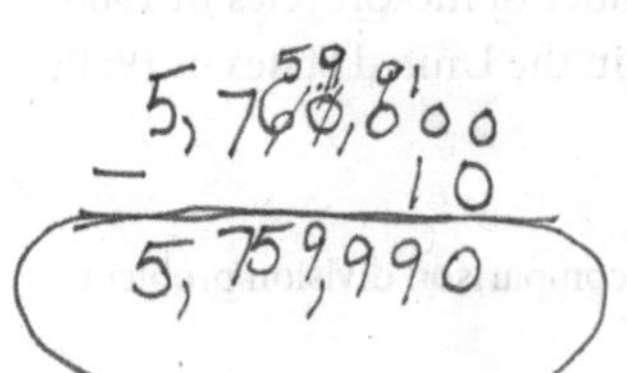

Figure 7.21 Jaylen's response to a multiplicative comparison division problem.

576,000
10 ⟌ 5,760,000
50
76
70
60
60
00
There was a total of 576,000 motor cycles in 1960.

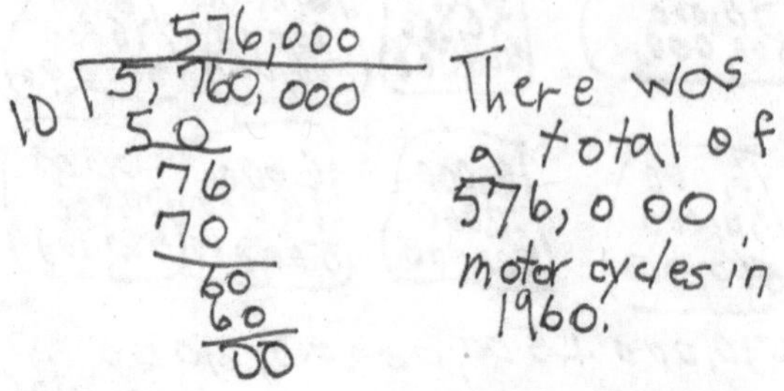

5. **Interpreting Remainders in Division Problems**: As described in this chapter, the contextual situation of a division problem determines how remainders should be interpreted. Study each of the following problems. For each, determine if there is a remainder, how the remainder should be interpreted, and the features of the problem that affected your decision.

   Problem 1: Five friends earn money by mowing lawns in their neighborhood. They evenly divide the money they earn each month. Last month the friends earned $125.00. How much did each friend receive?

   Problem 2: An average 4-year-old child is about 3½ feet tall. A typical giraffe is about 19 feet tall. About how many times taller is a typical giraffe than an average 4-year-old child?

   Problem 3: The fifth grade class celebrated the last day of school with a pizza lunch. The 24 students in the class equally shared 64 pieces of pizza. How many pieces of pizza did each student receive?

   Problem 4: A school parent group donated 50 markers to be shared equally among the 18 kindergarten students. How many markers should each kindergarten student receive?

   Problem 5: One hundred and twenty-three players signed up for the basketball league. The league organizers plan to create teams with 8 players per team. How many teams should the league organizers create?

6. **Try This in Your Class**: Follow the three steps to gather evidence of how your students are conceptualizing division.
   - Design or select a division question based on the mathematical goal of an upcoming lesson. Consider the different types of division problems you learned about in this chapter, as well as particular problem structures that can provide the type of information most helpful to you.
   - Administer the question as an exit question at the end of the lesson.
   - Analyze your students' responses and record the information on a copy of the *OGAP Evidence Collection Sheet* shown in Figure 2.24.
   (a) Use the evidence you collected earlier to answer the following three questions:
      - What are some developing understandings you noticed in the solutions that can be built upon in future lessons?
      - What are some underlying issues or concerns across your class that future lessons should address?
      - What are some implications for instruction? What are some specific instructional actions you can take to address the evidence you collected?

Instructional Link

Use the following questions and Table 5.2 to help you think about the ways in which your instruction and math program provide students multiple opportunities to engage in a range of problem contexts and structures involving division.

1. Do you or your instructional materials provide opportunities for students to:
   (a) Solve a range of partitive and quotative division across different problem contexts?
   (b) Solve a range of problems differing in contexts and structures that require students to interpret remainders?
2. To what degree does your math program focus on developing understanding and fluency for division through *Transitional Strategies* such as:
   (a) Visual models?
   (b) Place value understanding?
   (c) Properties of operations?
3. To what degree do you collect and analyze evidence in student solutions to inform instruction and student learning?
4. Based on your responses in Questions 1, 2, and 3 identify modifications you can make in your instruction to help your students gain a deeper conceptual understanding of whole number division.

8

Understanding Algorithms

Big Ideas

- The open area model builds conceptual understanding of both the distributive property and place value that can help students make sense of multidigit multiplication and division.
- Students should have opportunities to build explicit connections between the open area model and multiplication and division algorithms in order to build procedural fluency with understanding.
- Over time, as conceptual understanding is built and deepened, students should move away from models and less efficient strategies toward generalizable and efficient algorithms for multiplication and division.

Researchers have consistently indicated that students may struggle with the use and understanding of formal algorithms when their knowledge is dependent primarily on memory rather than anchored with a deeper understanding of the foundational concepts. Understanding and procedural fluency should be built in a way that brings meaning to both (Hiebert & Carpenter, 1992; Lampert, 1986; NRC, 2001; Russell, 2000). This chapter focuses on building procedural fluency with understanding through exploring the relationships among the open area model, properties of operations, and different algorithms for multiplication and division.

The importance of the open area model for developing understanding of multiplication and division and the properties of these operations has been established in previous chapters. On the *OGAP Multiplication Progression*, the use of the open area model is considered a *Transitional* strategy. As students' understanding of the operations and related concepts such as place value develops over time, it is important that they develop strategies that are increasingly efficient and generalizable, as represented at the *Multiplicative* level of the progression. Anchoring procedures to visual models is key to this process and involves several of the *Standards for Mathematical Practice* (CCSSO, 2010).

Go To See Chapters 3: The Role of Visual Models, Chapter 4: The Role of Concepts and Properties, and Chapter 7: Developing Whole Number Division for more information about using area models in the development of fluency.

In mathematics an algorithm is a procedure or series of efficient steps for solving a problem. Many different algorithms for multiplication and division are used across the United States and across different countries and cultures (Fuson & Beckmann, 2012). Figure 8.1 shows two multiplication algorithms and two division algorithms that are commonly presented in US mathematics textbooks. There are variations in how the steps of these algorithms are written, but here we show the most common recording methods. You will notice that these algorithms appear at the *Multiplicative* level, along with other strategies that are based on recall or properties of operations, on the *OGAP Multiplication and Division Progressions*.

Figure 8.1 Standard algorithms for multiplication and division.

Traditional US Multiplication Algorithm

$$\begin{array}{r} \overset{5}{5}7 \\ \times\ \ 8 \\ \hline 456 \end{array}$$

Partial Products Algorithm

$$\begin{array}{r} 57 \\ \times\ \ 8 \\ \hline 400 \\ 56 \\ \hline 456 \end{array}$$

Traditional US Division Algorithm

$$\begin{array}{r} 57 \\ 8\overline{)456} \\ -40 \\ \hline 56 \\ -56 \\ \hline 0 \end{array}$$

Partial Quotients Algorithm

$$\begin{array}{r} 57 \\ 8\overline{)456} \\ -400\ \ \ 50 \\ \hline 56 \\ -56\ \ \ \ 7 \\ \hline 0 \end{array}$$

These are all examples of standard algorithms that are efficient, accurate, and generalizable for all whole numbers and decimals. In this book, we refer to the multiplication and division algorithms on the left side of Figure 8.1 as the *traditional US algorithms* because they are more widely used in this country, but it is important to note that we are not suggesting they are preferred methods. More importantly all algorithms should be built upon a strong foundation of place value understanding, properties of operations, and connections to visual models in order to develop procedural fluency with understanding.

CCSSM

In the CCSSM, fluency with standard algorithms for multiplication and division is not expected until students have developed conceptual understanding of strategies based on place value and properties of operations. Table 8.1 shows the progression of models, strategies, and algorithms that are expected from grade 2 to grade 6 in the domains of Operations and Algebraic Thinking and Numbers and Operations in Base 10.

Note the important role of strategies based on visual models, place value, properties of operations, and inverse relationships between operations during the two years before students are expected to be able to use the standard algorithms. In addition, the standards include the expectation that students should be able to "illustrate and explain the calculation by using equations, rectangular arrays, and/or area models" for both multiplication and division before fluency with a standard algorithm is expected. Thus, the CCSSM also reflects a progression from the use of visual models to develop conceptual understanding to the use of standard procedures without models.

Table 8.1 Progression of multiplication and division models and strategies in the CCSSM.

Grade	2	3	4	5	6
Use of Models	Equal groups Arrays	Equal groups Arrays Area	Arrays Open area	Volume Arrays Open area	
Multiplication Strategies	Repeated addition	Strategies based on place value and properties of operations		Standard algorithm	
Division Strategies			Strategies based on place value, properties of operations, and/or the inverse relationship between multiplication and division	Standard algorithm	

(handwritten note in table: use inverse relationship and ~~ strategies)

The CCSSM does not provide a specific definition of standard algorithms, but in the *Progression for the Common Core State Standards in Mathematics* for the domain of Number and Operations in Base 10, the following definition is offered:

Standard algorithms for base-ten computations with the four operations rely on decomposing numbers written in base-ten notation into base-ten units. The properties of operations then allow any multi-digit computation to be reduced to a collection of single digit computations. These single-digit computations sometimes require the composition or decomposition of a base-ten unit (Common Core Standards Writing Team, 2015, p. 13).

The authors of the progression go on to explain that the algorithm is "defined by its steps" and that "minor variations in recording standard algorithms are acceptable" (p. 13).

Later in the chapter, we will consider the implications of this defining characteristic that standard algorithms can be "reduced to a collection of single digit computations" for developing procedural fluency with understanding, but first we explore the importance of developing understanding of multidigit multiplication and division through visual models.

Multiplication: Linking Open Area Models and Algorithms

In order to explore the meaning of multidigit multiplication more deeply, try this: solve a two-digit by two-digit multiplication problem in three different ways— using open area, partial products, and the traditional US algorithm—and then explicitly try to draw connections between the strategies. As you carry out the steps in each strategy, try to think about the mathematics behind the steps you are carrying out, the words you say to yourself as you carry them out, and the place value of those quantities in the original problem (See Looking Back #1).

To further explore the connections among the open area model, the partial products algorithm, and the traditional US algorithm we will look more closely at student solutions to an equal measures problem that involves multiplying two-digit quantities.

Open Area and Partial Products Algorithm

Look at Alyssa's and Gavin's solutions to the problem shown in Figure 8.2. What connections can you find between the open area model and Gavin's use of partial products in Figure 8.2? Do you see some of the same numbers? Where? Where do these numbers come from?

Figure 8.2 Comparing open area and partial product solutions. The four partial areas are equivalent to the four partial products.

Twenty-three inches of string are needed to hang each decoration. How many inches of string are needed to hang 27 decorations?

Alyssa's Solution

Gavin's Solution

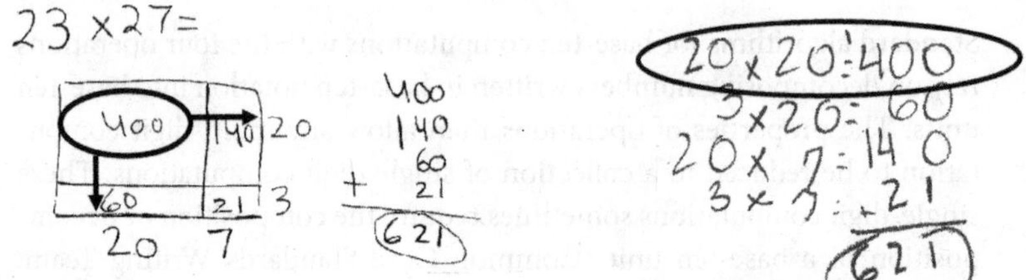

Note that in Alyssa's open area model, the dimensions of the partial areas are the factors of the four partial products in Gavin's solution ($20 \times 20 = 400$, $3 \times 20 = 60$, $20 \times 7 = 140$, and $3 \times 7 = 21$). In Figure 8.2, the product of $20 \times 20 = 400$ is circled in both strategies. Can you identify the other three partial products in Alyssa's model?

Open Area and Traditional US Algorithm

Now compare Kayla's use of the traditional US algorithm in relation to Alyssa's open area model in Figure 8.3.

Figure 8.3 Comparing open area and traditional US algorithm. Two partial products are added together in the traditional US algorithm.

Twenty-three inches of string are needed to hang each decoration. How many inches of string are needed to hang 27 decorations?

Alyssa's Solution   Kayla's Solution

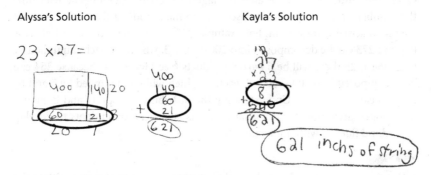

The partial products in the traditional US algorithm are a little harder to find because they have been combined to create the two partial products of 81 and 540. That is, 81 comes from $27 \times 3 = (7 + 20) \times 3 = (7 \times 3) + (20 \times 3) = 21 + 60$. These are also the products on the bottom row of Alyssa's open area model (60 and 21) as shown in Figure 8.3. Likewise, the 540 in Kayla's solution comes from $27 \times 20 = (7 + 20) \times 20 = (7 \times 20) + (20 \times 20) = 140 + 400$. Can you identify these partial products in Alyssa's open area model?

Recognizing that each of the numbers that are produced in the traditional US multiplication algorithm is a sum of the partial products can help bring meaning to the standard algorithm. In this way, the open area model helps bring meaning to multidigit multiplication and can eventually become a mental model to support sense making. Researchers indicate that fluency and understanding be built using visual models and the mathematical concepts underpinning the operations rather than just rote memorization of algorithmic steps (e.g., Battista, 2012; Carpenter et al., 2003; Empson & Levi, 2011; Fosnot & Dolk, 2001; Kaput, 1989).

Note that in her open area model, Alyssa has drawn the dimensions proportionally (20 is longer than both 7 and 3). Her solution shows how the product of 20 and 20 make up the largest proportion of the total area and provides a rough estimate of the magnitude of the answer to 23 × 27. It is important to encourage students to continually make sense of the open area model by asking questions such as:

- What multiplication problem does your open area model represent?
- What does this part of the area model represent? How do you know? Can you write an equation?
- How did you decompose this number? Why?
- Which part can give you an estimate of the total answer?
- Why do you add, not multiply, the partial products?
- Why does this answer make (or not make) sense?

Over time, students should be able to imagine the open area model to determine the number of partial products based on the magnitude of the factors without having to actually draw it out. For example, 273 × 6 will have three partial areas because 273 can be decomposed into 200 + 70 + 3, which are each multiplied by 6. In 364 × 25 there will be 6 partial products (see Figure 8.4) because 264 can be decomposed into 300 + 60 + 4, each of which will be multiplied by 5 and by 20 because 25 = 20 + 5. Understanding the distributive property and linking it to the open area model and partial products is critical for students to develop procedural fluency with understanding.

 Go To See Chapter 4: The Role of Concepts and Properties for more on the distributive property.

Figure 8.4 Open area model for 364 × 25.

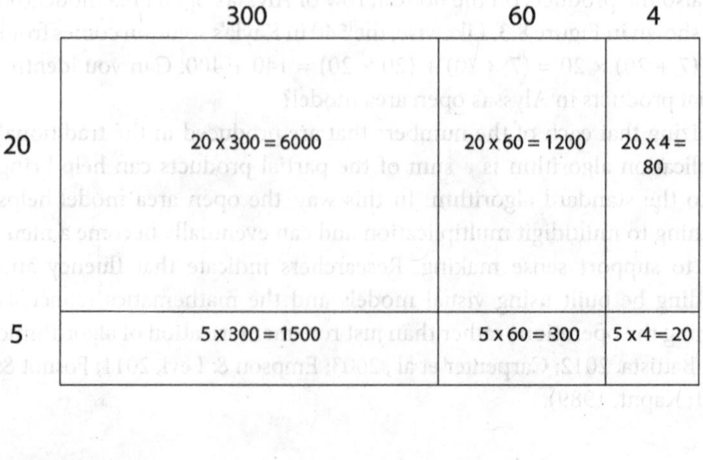

	300	60	4
20	20 × 300 = 6000	20 × 60 = 1200	20 × 4 = 80
5	5 × 300 = 1500	5 × 60 = 300	5 × 4 = 20

From Models to Standard Algorithm

The partial product algorithm is easily connected to the open area model. Note that in Alyssa's solution in Figures 8.2–8.3 she has written the four partial products next to the open area model. A logical next step for her would be to introduce the partial products method and show that it is simply writing out the products without drawing the model. In addition to being easy to connect to the open area model, the partial products algorithm has several advantages. In each step, the place value of the quantities is kept intact, forcing students to work with place value and develop deeper understanding of multidigit quantities. Unlike the traditional US algorithm, all the multiplication is done first and then the addition. Furthermore, students can work from right to left or left to right, which for some students may seem more natural because they learn to read both numbers and text from left to right (Fuson & Beckmann, 2012). When working from the left to right, the partial products algorithm starts with the largest, rather than the smallest, place value of each factor, providing students with a rough estimation of the magnitude of the answer after the first step.

The traditional US algorithm also has advantages: it is efficient and concise for multidigit calculations and can be generalized across all whole numbers and decimal quantities. Historically, it was very useful for doing multidigit arithmetic before calculators and computers were available because it reduces difficult calculations to a series of one-digit products and sums (Fosnot & Dolk, 2001).

Fuson and Beckmann (2012) argue that this generalizability applies to variations such as partial products, which can also be considered a standard algorithm: "The standard algorithms are especially powerful because they make essential use of the uniformity of the base-ten structure. This results in a set of iterative steps that allow the algorithm to be used for larger numbers" (p. 16). Exploring and recognizing this uniformity and regularity allows students to engage in two important mathematical practices described in the CCSSM: "Look for and make use of structure" and "Look for and express regularity in repeated reasoning" (CCSSO, 2010).

However, pedagogically standard algorithms are not as useful when introduced before students have developed strong place value understanding. The very thing that makes the traditional US algorithm so efficient—the reduction to single-digit calculation—also ends up masking the place value or meaning of the quantities. It also makes it difficult to carry out mentally or make an estimate of the product. By combining two operations in one step (two partial products are calculated and then added), it is harder for students to make sense of the place value of the quantities being operated on and the underlying mathematics of the operation. By reducing the calculation to a series of one-digit multiplications, the arithmetic becomes easier, but students are not required to draw on or develop place value understanding or operate on groups of 10. In fact, premature introduction of algorithms that mask place value can constrain student opportunities to reflect on the tens-structure of the number system (Ebby, 2005; Kamii, 1998).

Try this: Compute 27 × 38 using the traditional US algorithm (see Figure 8.5). Talk through each step and note how many times you say something that is not mathematically true. Table 8.2 shows that only the first and last steps are true to the mathematical meaning of the numbers being multiplied.

This activity shows that, in essence, students need to leave their number sense aside when carrying out this algorithm. It is not so surprising, then, that students often make errors when they learn standard algorithms before developing understanding.

Figure 8.5 Solving 27 × 38 with the traditional US algorithm.

$$
\begin{array}{r}
\overset{2}{\underset{5}{}} \\
27 \\
\times\ 38 \\
\hline
216 \\
810 \\
\hline
1026
\end{array}
$$

Table 8.2 Deconstructing the steps of the traditional US algorithm for 27 × 38 in relation to the mathematical meaning.

What is Said:	Not True	The Mathematical Meaning
7 times 8 is 56		
Put down the 6		The 6 is in the one's place of the product, 56
Carry the 5	X	The 5 is really 50 from 56
8 times 2 is 16	X	8 × 20 is what you are solving and it is 160
Plus 5 is 21	X	The 5 really means 50 and 160 + 50 = 210
Put a zero (or an X) to hold the place	X	You are multiplying by a multiple of a power of 10 (30) so the product will end in a zero
3 times 7 is 21	X	30 × 7 is what is being multiplied
Put down the 1	X	The 1 is a 10 because 30 × 7 = 210
Carry the 2	X	30 × 7 = 210 and the 2 represents 200
3 times 2 is 6	X	This is really 30 × 20 = 600
Plus 2 is 8	X	The 2 is 200 and the 8 is 800
Add 216 and 810		

Algorithms Without Understanding

The compactness of standard algorithms makes them very useful but at the same time makes them prone to procedural use without understanding. As Bass (2003) notes, this is particularly problematic for traditional algorithms that have opaque steps:

> Traditional algorithms have evolved over time for frequent daily users who want to do routine calculations, essentially mechanically. They tend to be cleverly efficient (minimizing the amount of space and writing used) but also opaque (the steps are not notationally expressive of their mathematical meaning). Therefore, if these algorithms are learned mechanically and by rote, the opaque knowledge, unsupported by sense making and understanding, often is fragile and error-prone, as many researchers have documented (p. 326).

In the examples shown in Figure 8.6, students are solving a multidigit equal groups problem by attempting to use the traditional US algorithm. Both students started out with the correct first step (3 × 6 = 18) and seemed to be able to use some parts of the procedures. The errors seem to crop up when students have to combine the multiplication and addition in one step.

Figure 8.6 Common errors when using the US traditional algorithm.

The Arbor Tree farm planted 6 fields with 273 trees in each field. How many trees did they plant in all?

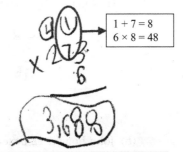

Multiplies 6 × 3 to get 18 and 6 × 7 to get 42. Adds the 1 from 10 to get 43, but then records the 3 and carries a 1 instead of a 4 from 43. Perhaps an overgeneralization that you always carry a one.	Multiplies 6 × 3 to get 18, but then adds the 1 (from the 10 in 18) to the 7 and then multiplies 6 × 8 to get 48. Makes the same error in adding the 4 to the 2 and then multiplying 6 × 6 to get 36. Adds then multiplies instead of multiplying and then adding.

Both students are correctly following some of the steps of the traditional US algorithm but are not able to put all of the steps together. In Solution A, the student may have overgeneralized the idea that a 1 is always carried. In Solution B, the student may be confused about why the operations of addition and multiplication are combined in one step. Although some might argue that these are careless errors, the evidence suggests that the students are not using the procedure with understanding. Solution B also has an unreasonable answer. Both solutions would be considered *Nonmultiplicative* on the *OGAP Multiplication Progression*—the students are attempting to use a multiplicative strategy but are not demonstrating underlying multiplicative thinking. These students would most likely benefit from going back to the open area model to conceptualize the three partial products and develop the understanding that 200, 70, and 3 are all multiplied by a factor of 6.

In Figure 8.7, Jesse is able to solve the first question by using the traditional US algorithm to multiply a 2-digit by a 1-digit number. However, in the second part, which involves multiplying two 2-digit quantities, the fragility of his understanding becomes evident. (Some might say he just forgot to put in the "place holder," which highlights the procedural nature of the use of this algorithm. What exactly is a place holder?) More importantly, he doesn't seem to

Figure 8.7 Limits of procedural understanding. Jesse uses the traditional US algorithm to solve multiplication problems, but runs into problems when there are two-digit factors.

(a) Mark bought 12 boxes of crayons. Each box contained 8 crayons. How many crayons were there all together? Show your work.

$$\overset{1}{8}$$
$$\times 12$$
$$\overline{98\,\text{Crayons}}$$

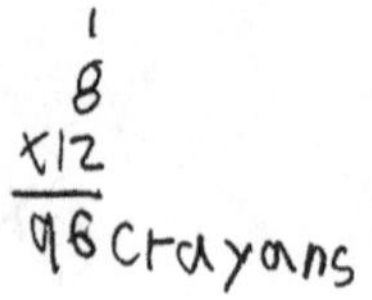

(b) John bought 12 boxes of crayons. Each box contained 64 crayons. How many crayons were there all together? Show your work.

$$\overset{1}{12}$$
$$\times 64$$
$$\overline{\overset{1}{4}8}$$
$$+72$$
$$\overline{120}$$

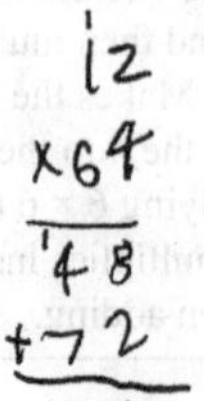

notice his answer is unreasonable given the factors in the problem. Open area model and partial products could be used to help Jesse see the second product as 720 rather than 72 and recognize that the answer to 12 × 63 cannot possibly be 120, because 12 × 10 is 120. Jesse's work on the second part of this problem would be considered *Nonmultiplicative* on the *OGAP Multiplication Progression* because he is using a procedure incorrectly. More importantly, his use of the algorithm is not anchored to conceptual understanding of place value and properties of operations.

It may be the case that Jesse was introduced to the traditional US algorithm too early, before he developed understanding of place value and multiplication by multidigit numbers. Students are also apt to make errors if the partial products algorithm is taught procedurally, and they need to have solid understanding of place value in order to use the procedure effectively. In the example shown in Figure 8.8, Jasmine has multiplied 200, 70, and 3 by 6 to find the partial products, but has made a place value error in computing 200 × 6 as 12,000. However, because the place value of the quantities and the steps to the procedure are transparent in this algorithm, it is easier for a teacher to identify and remedy these errors by making connections back to the open area model.

Figure 8.8 Errors with the partial products algorithm. Jasmine's work shows a place value error.

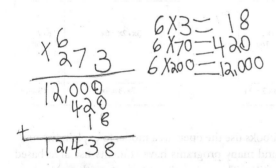

In sum, both the traditional US algorithm and the partial products algorithm can be taught and learned as rote procedures, which make them "highly dependent on memory and subject to deterioration" (Kieren, 1988, p. 178). However, if students have developed strong and deep understanding of the operations, the properties of multiplication and division, and place value through open area models, they are better positioned to understand the meaning behind the steps of the standard multiplication and division algorithms and can also appreciate the generality these algorithms afford: "Generality is one of the most important and powerful characteristics of mathematics" (Bass, 2003, p. 326).

Algebraic Connections

This generality can be seen in the fact that the partial products algorithm and the open area model are related to how one can multiply algebraic expressions or polynomials. A *polynomial* is an algebraic expression consisting of variables and coefficients, such as $5x^2 + 4x + 7$. A polynomial with only two terms, such as $6x + 8$, is called a *binomial*. The method for multiplying binomials that many people will remember from algebra class relies on the distributive property and is sometimes referred to as FOIL for "first, outer, inner, last." These four terms signify the four partial products: For $(3x + 8)(2x + 3)$ one multiplies the *first* terms of each binomial ($3x \cdot 2x = 6x^2$), then the *outer* terms—first term of the first binomial and second term of the second binomial ($3x \cdot 3 = 9x$), then the *inner* terms—the second term of the first binomial with the first term of the second binomial ($8 \cdot 2x = 16x$), and then finally the *last* term of each binomial ($8 \cdot 3 = 24$). The result is $6x^2 + 9x + 16x + 24$ or $6x^2 + 25x + 24$. Study the diagram in Figure 8.9 which highlights the connections between this method and the open area model. If one thinks of a two-digit quantity such as 38 in expanded form, as $(3 \times 10) + 8$, the connections to $3x + 8$ become apparent.

Figure 8.9 Using the open area model to see connections between whole number and binomial multiplication.

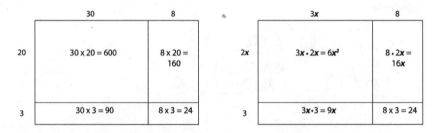

In fact, most algebra textbooks use the open area model to explain the multiplication of polynomials, and many programs have students use area-based manipulatives called algebra tiles. Thus, providing students with experiences to make sense of the open area model in grades 3–5 for multiplication can help them make stronger connections and develop meaning for the algebraic manipulations they will use and learn later on in mathematics.

Division Algorithms

The open area model can also be used to develop understanding of the inverse relationship between multiplication and division and help students make sense of division and division algorithms. (See Chapter 2 for more on the area model and Chapter 7 for more on division.) Because division is a missing factor problem, finding the quotient can be thought about as finding one dimension of a rectangle when the area and the other dimension are known. In Figure 8.10, the problem 852 ÷ 6 is shown as a rectangle with area of 852 and one dimension of 6. The other dimension is the unknown. In this case, because the product is much greater than the dividend of 6, we know we are finding a much longer dimension. Building on the open area model to make sense of division can also help students make sense of the standard symbol separating the dividend from the divisor in the standard division algorithms.

Figure 8.10 Understanding division with the open area model.

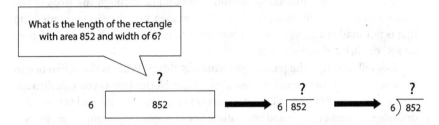

The open area model can be used in this way to help students make sense of division, but it is important to note that it is not necessarily an effective method for students to use to solve division problems. In Figure 8.11, the steps of the partial quotients and traditional US long division algorithm can be connected to the area model to help students make sense of steps of the algorithms that otherwise might seem opaque. This is particularly important in the long division algorithm where the place values of the quantities are not represented.

Figure 8.11 Connecting division algorithms with the open area model.

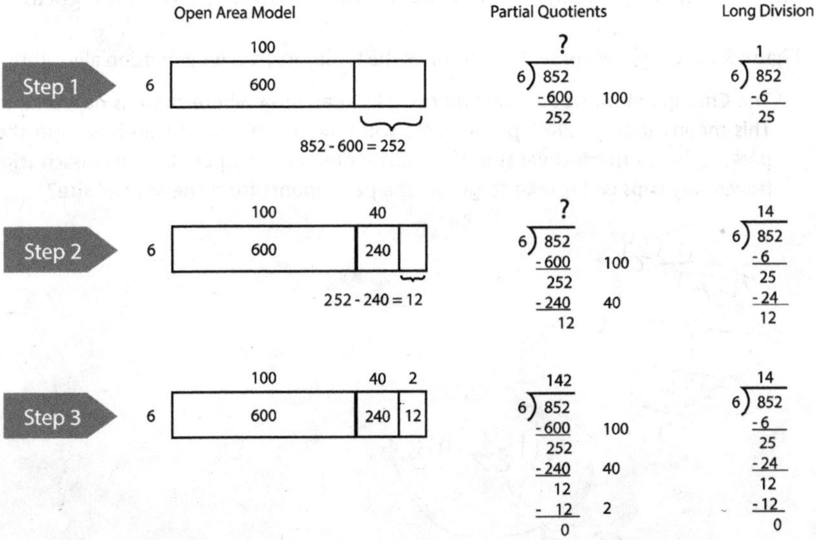

As Figure 8.11 illustrates, the partial areas in the open area model are also the partial quotients in the partial quotient algorithm. One can think "how many 6's are in 852" and begin by taking out 100 groups of 6 because the product of $6 \times 100 = 600$. Because 600 is only part of the total area, one then subtracts to find that 252 units remain. In the second step, one takes out 40 additional groups of 6, or 240. There are then 12 units left, or 2 groups of 6 (step 3).

This is also parallel to the steps of the traditional US long division algorithm, but here the quantities are again reduced to single-digit computations (6×1, 6×4, and 6×2 rather than 6×100, 6×40, and 6×2). Try the same exercise from Table 8.2 with these division algorithms. As you talk through the steps of the traditional US long division algorithm, how many times do you say something that is not mathematically true ("1×6 is 6") or mathematically vague ("6 goes into 8" or "bring down the 5")?

Now talk through the partial quotients algorithm. What is the difference in the language you use in each of these algorithms? Notice that as you talk through the partial quotients steps all the statements you make are true, and you are also drawing on number sense and estimation. For example, how many 6's are in 852? There are at least 100 because $100 \times 6 = 600$, and 852 minus 600 equals 252. How many 6's are in 252? At least 40 because $40 \times 6 = 240$, and 252 minus 240 is 12, so there are two more 6's because 6×2 is 12. Every statement is mathematically true and maintains the meaning of the quantity (i.e., the base-10 number system). In addition, all the computations involve multiples of 10, 100, etc.

Although the traditional US long division algorithm may seem easier because of the reduction to one-digit computations, it can be hard for students to keep track of the place value of the quantities, particularly when there are zeros in the dividend or when the divisor has multiple digits. In the example shown in Figure 8.12, Cary attempts to use the traditional US long division algorithm

Figure 8.12 Cary's response. Cary misuses the traditional US long division algorithm.

The Champlain Music Festival will be held in an area where there is no parking. This means that all 2881 participants will have to take a shuttle bus from the parking lot to the festival site. If a shuttle bus can transport 67 riders each trip, how many trips will it take to get all the participants from the festival site?

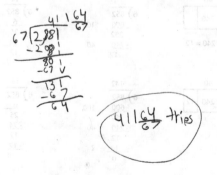

to solve a quotative division problem. His miscalculation that 67 × 4 equals 208 rather than 268 resulted in needing to subtract 67 twice, which he has recorded with two 1's in the quotient, resulting in an unreasonable answer of 411. In this case, he seems to be following the steps of the traditional US long division algorithm without connecting those steps to an understanding of division, or the process of taking out equal groups of 67. He also misinterprets the remainder, as there cannot be a fraction of a trip taken. (See Chapter 7: Developing Whole Number Division for more on interpreting remainders.)

On the other hand, if students are fluent with multiplying by powers of 10 and multiples of powers of 10 (see Chapter 4), the partial quotients algorithm does not require difficult calculations and keeps the place value of the quantities intact. Compare the following two solutions to the same quotative division problem in Figures 8.13 and 8.14.

Figure 8.13 Bianca's response. Bianca uses the traditional US long division algorithm.

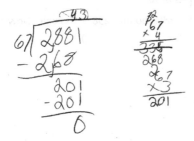

Figure 8.14 Alex's response. Alex uses the partial quotient algorithm to repeatedly subtract multiples of 10.

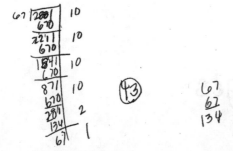

In Bianca's solution in Figure 8.13, it appears she may have multiplied 67 by 5, then 3, and then finally 4 using the traditional US multiplication algorithm in order to find the appropriate product to start with in the first step of the long division algorithm. In Figure 8.14 Alex, on the other hand, began with an easy product (67 × 10) and then repeatedly took that product out of the quotient. An advantage of the partial products algorithm is that students can work with products they can calculate mentally if they have a strong foundation in working with multiples of powers of 10. Note, however, that Alex's solution involves a lot of steps and a lot of subtraction.

Study the two solutions in Figures 8.15 and 8.16. Both students used the partial quotients algorithm, but in Figure 8.16 Savannah began with a larger product (3 × 80 = 240), making the algorithm more efficient and potentially more accurate. Both Alex's (Figure 8.14) and Nasir's (Figure 8.15) less efficient approaches are considered *Transitional Strategies* on the *OGAP Division Progression*.

Figure 8.15 *Transitional Strategy*. Nasir's solution shows evidence of inefficient partial quotients by repeatedly subtracting multiples of 10.

Tennis balls come in packages of 3. Jordan bought 258 tennis balls. How many packages did he buy?

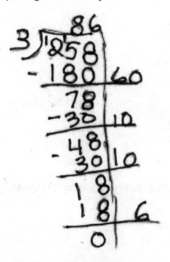

86 packages

Figure 8.16 *Multiplicative Strategy.* Savannah's use of the partial quotient algorithm is efficient as she starts by taking out the largest multiple of 10 possible.

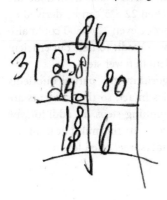

Helping students develop more efficient strategies is an important part of moving from the *Transitional* to the *Multiplicative* stage on *the OGAP Multiplication and Division Progressions* (see Chapter 2). One way to do this with division is to have students brainstorm all the products they can mentally calculate with the divisor in order to find the most efficient partial quotient. In the example shown in Figure 8.17, Quadir has created a list, or *menu*, of products of 22 before starting to divide 341 by 22.

Figure 8.17 Quadir's response. Quadir first creates a menu of products of 22 before choosing which one to start with for the partial quotients algorithm.

The area of a rectangular garden is 341 square feet. If the length is 22 ft., what is the width of the garden?

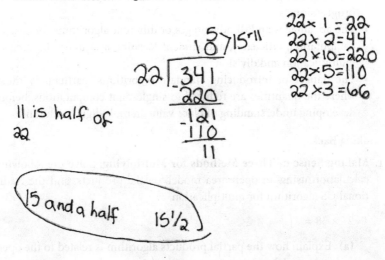

These products can be generated easily by using relationships between the factors. For example, 22×2 and 22×10 are easily derived facts from doubling and multiplying by powers of 10, respectively. Then 22×5 can be derived by halving the product of 22×10. (22×15, 22×30, or even 22×500 could also be easily derived from this list). The menu allows Quadir to take out efficient partial quotients of 10 and 5 and solve the problem in fewer steps.

Unlike the traditional US long division algorithm, the partial quotients algorithm keeps the place value of the quantities intact, so that students are estimating the value of the quotient in the beginning steps. In addition, by working with powers of 10 and multiples of powers of 10, the multiplication and subtraction calculations are not difficult or cumbersome, often leading to less frustration on the part of the student.

Chapter Summary

As the authors of the *Number and Operations in Base 10 Progression* state, "By reasoning repeatedly about the connection between math drawings and written numerical work, students can come to see multiplication and division algorithms as abbreviations or summaries of their reasoning about quantities" (CCSSM writing team, 2015, p. 14). In the *OGAP Multiplication and Division Progressions*, standard algorithms can be found at the *Multiplicative* level, but they should be built upon a strong foundation on and connection to visual models in the *Transitional* level.

This chapter focused on:

- Developing procedural fluency with understanding by connecting algorithms to place value, visual models, and properties of operations
- The importance of developing foundations of conceptual understanding along with efficiency as student strategies move along the progression
- The advantages and disadvantages of different algorithms for ease of computation, efficiency, and student learning and understanding of multiplication and division
- The danger of introducing standard algorithms, particularly those where the quantities are reduced to single-digit computations, before developing understanding of place value in multidigit operations

Looking Back

1. **Making Sense of Three Methods for Multiplying:** Solve the following calculation using an open area model, partial products, and the traditional US algorithm for multiplication.

    $27 \times 38 =$

    (a) Explain how the partial products algorithm is related to the open area model.

(b) Explain how the traditional US algorithm is related to the open area model.

(c) Explain how the open area model and partial products algorithm can be used to solve mixed-number multiplication problems such as $1\frac{1}{2} \times 1\frac{1}{4}$.

(d) Explain how the open area model is related to multiplying algebraic expressions like $(2x + 3)(3x + 4)$.

2. **Examining the Traditional Algorithm for Division**: Study the example of the traditional algorithm for division shown in Figure 8.18.

(a) Where in the algorithm is the answer to 3×12?

Figure 8.18 An example of the traditional US algorithm for division.

```
          273
    12) 3276
      - 24
         87
       - 84
         36
       - 36
          0
```

(b) Where in algorithm is the answer to 200×12? How do you know?

(c) Use the example to reason about the answer to 270×12 and then 203×12.

3. **Comparing Student Solutions:** Kira's and Jovan's solutions to the crayon problem are shown in Figures 8.19 and 8.20.

The Crayon Problem

John bought 12 boxes of crayons. Each box contained 64 crayons. How many crayons were there all together? Show your work.

Figure 8.19 Kira's solution.

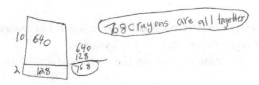

Figure 8.20 Jovan's solution.

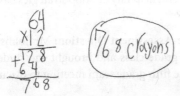

    (a) Why does Kira's open area model have only two partial products rather than four as one might expect for a two-digit by two-digit calculation?

    (b) How are Kira's and Jovan's solutions alike? How are they different?

    (c) What questions could you ask both students to ensure they have conceptual understanding of the algorithms they are using?

    (d) Imagine you have both Kira and Jovan share their solutions with the class. What questions could you ask the class to help them see the connections between the two strategies?

4. **Analyzing a Student Solution:** Ms. Bright brought the student work shown in Figure 8.21 to share with her colleagues at a team meeting.

Figure 8.21 Lily's solution to an area problem.

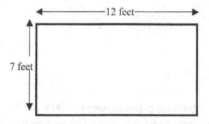

How many *one foot square tiles* are needed to tile the floor?

Show your work.

$$\begin{array}{r} \overset{1}{1}2 \\ \times\,7 \\ \hline 714 \end{array}$$

Answer
714 1 foot
Tiles.

    (a) What is the evidence of developing understanding in Lily's work?

    (b) What issues or concerns are there in Lily's solution?

    (c) Where would this solution fall on the *OGAP Multiplication Progression*?

    (d) What instructional strategies or models could Ms. Bright use to help address these concerns and help Lily develop strategies based on multiplicative reasoning?

5. **Analyze Kamal's Solution to a Pre-Assessment Question:** Mr. Johnson gave a pre-assessment to his fifth grade class and brought the student solution shown in Figure 8.22 to the fifth grade team meeting to discuss.

Figure 8.22 Kamal's solution to an area division problem.

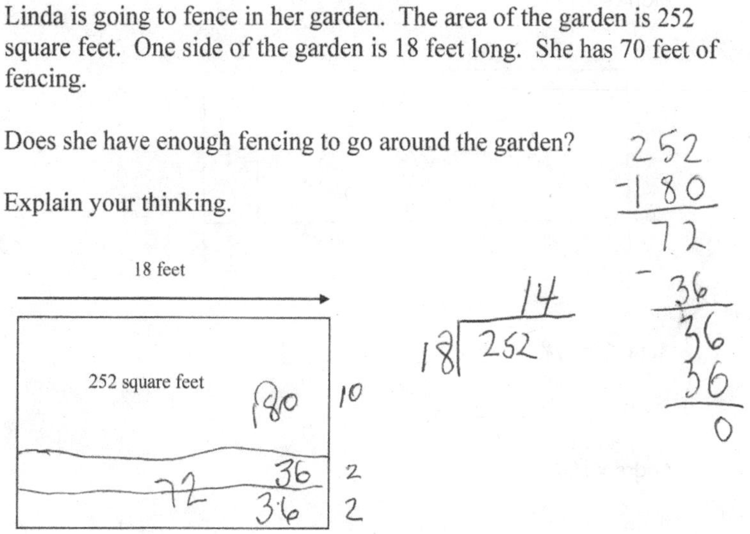

Linda is going to fence in her garden. The area of the garden is 252 square feet. One side of the garden is 18 feet long. She has 70 feet of fencing.

Does she have enough fencing to go around the garden?

Explain your thinking.

(a) What is the evidence of developing understanding in Kamal's work?

(b) Are there any issues or concerns in Kamal's solution?

(c) Where does Kamal's solution fall on the *OGAP Multiplication Progression*?

(d) How is Kamal's solution related to the partial quotients algorithm?

(e) What instructional moves could Mr. Johnson make to help him develop more efficient division strategies?

6. **Student Responses to a Partial Products Problem:** Ms. Jackson's students were working with the open area model to solve multidigit multiplication problems. She gave the partial products problem as an exit slip at the end of a lesson. Study the three student solutions in Figure 8.23 and answer questions a–c.

(a) What are the developing understandings evidenced in each solution?

(b) What issues or concerns are evidenced in each solution?

(c) What are possible next instructional steps for each student based on the evidence in their work on this problem?

Partial Products Problem

Label the four rectangles with the multiplication equations that represent the partial products model. In the following spaces write and solve the equation that matches the model. Show your work.

Figure 8.23 Three student solutions to the partial products problem.

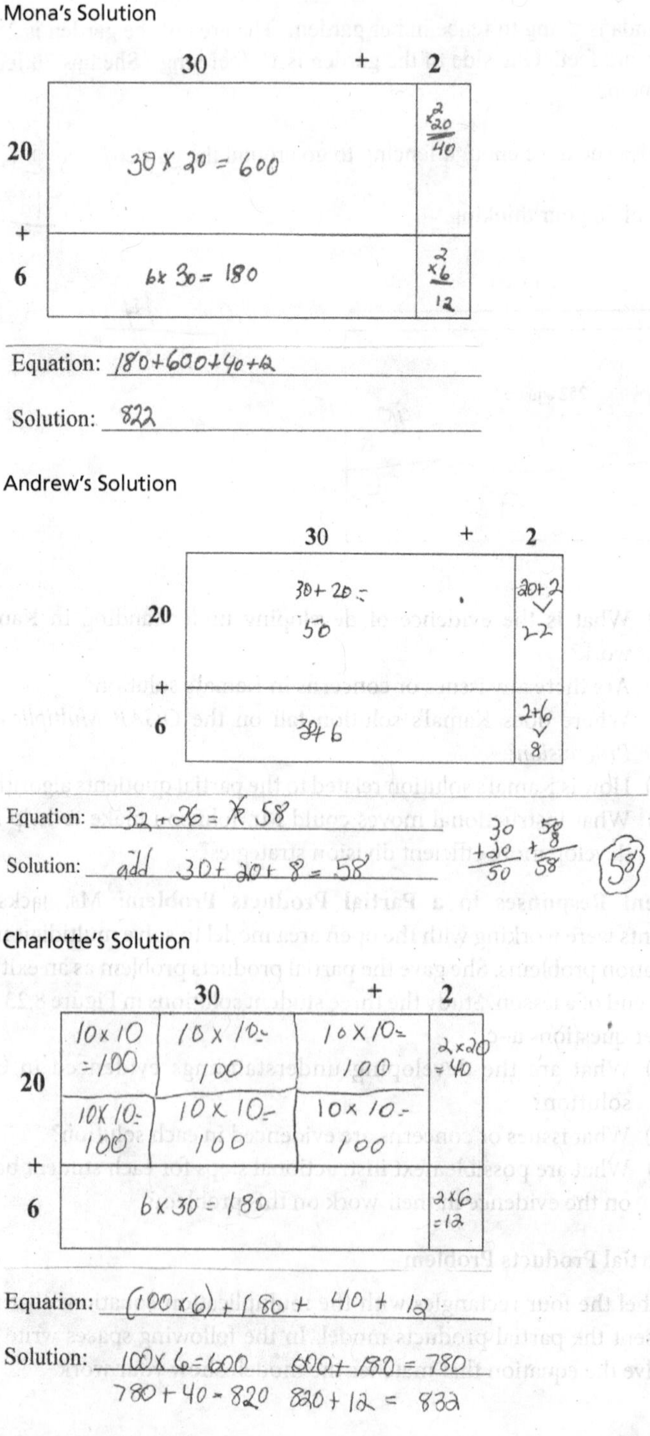

Mona's Solution

30 + 2

20	30 × 20 = 600	2 ×20 —— 40
+		
6	6 × 30 = 180	2 ×6 —— 12

Equation: 180+600+40+2

Solution: 822

Andrew's Solution

30 + 2

20	30+ 20 = 50	20+ 2 ∨ 22
+		
6	30+ 6	2+6 ∨ 8

Equation: 32 + 26 = ⟨ 58

Solution: add 30+ 20+ 8 = 58   30 58
+20 +8
—— ——
50 58   (58)

Charlotte's Solution

30 + 2

20	10×10 =100	10 × 10= 100	10× 10= 100	2×20 = 40
	10×10= 100	10 × 10= 100	10× 10= 100	
+				
6	6× 30 = 180			2×6 =12

Equation: (100×6) + 180 + 40 + 12

Solution: 100×6=600   600+ 180 = 780
780+40=820   820+12 = 832

Instructional Link

Use the following questions to help you think about how your instruction and math program provide students opportunities to develop procedural fluency with understanding for the algorithms appropriate for your grade level.

- What multiplication and division algorithms are introduced in your math program, and when are they introduced?
- Is there sufficient attention to developing understanding of the operations, foundational place value knowledge, the relationship between multiplication and division, and properties of operations before standard algorithms are introduced?
- Does your program draw explicit connections between algorithms, open area models, place value, and properties of operations?
- What are some ways to address any shortcomings or gaps you identified in your answers to the earlier questions?

9

Developing Math Fact Fluency

Big Ideas

- Development of concepts and strategies must come prior to drill, practice, and memorization of basic math facts.
- Students need multiple experiences developing strategies for finding products of basic math facts.
- Mastery of single-digit multiplication facts is critical for supporting flexibility and understanding of more complex mathematics concepts.

Any teacher of mathematics will tell you students need to have automatic recall of single-digit math facts. The debate lies, however, in how to achieve this goal. The importance of automaticity with basic facts becomes most compelling when students do not have easy access to them. While students struggle to recall facts, the focus of a lesson can be lost and the pace of a lesson may be slowed, drawing away from the goals of the lesson (Forbringer & Fahsl, 2009). Additionally, researchers indicate:

> Basic math facts are considered to be foundational for further advancement of mathematics. They form the foundation for learning multi-digit multiplication, fractions, ratios, division, and decimals. Many tasks across all domains of mathematics and across many subject areas call upon basic multiplication as a lower order of the overall task (Wong & Evans, 2007, p. 91).

Clearly learning the basic multiplication and division facts is important. The question is how best to support students in gaining automaticity and fluency with them. This chapter focuses on what it means to develop fluency and automaticity, strategies that build fluency of the facts through understanding, as well as a discussion on targeted and general practice to support the development of fluency.

Let us begin with what we mean by fluency and automaticity. It is generally agreed that fluency of basic facts means that students can recall the fact with relative ease, often requiring a mental computation to occur. Fluency may not be as quick as automaticity, but it results in an accurate and relatively efficient response. Automaticity is effortless recall within about 3 seconds (Van de Walle et al., 2012) without performing any mental computation. Automaticity is synonymous with "know from memory."

It is important to understand that automaticity is not the same as simply memorizing. Automaticity is reliant on instruction that is initially focused on fluency. One reason that memorizing facts is so challenging for students and discouraged by both cognitive neuroscientists and math education researchers is that it is not reasonable to memorize so many pieces of isolated knowledge. According to the cognitive neuroscientist Stanislaw Dehaene (2011), "If our brain fails to retain arithmetic facts, that is because the organization of human memory, unlike that of a computer, is associative: It weaves multiple links among disparate data" (p. 113).

Strategies such as counting all or adding a series of numbers, which are considered low-level strategies, to derive an answer to a single-digit math fact is not efficient and is prone to errors. Because deriving multiplication facts with low-level strategies can be more time consuming than it is for addition facts, it is even more imperative that students develop efficient methods that rely on multiplicative relationships. This idea is the foundation of what is meant by fluency. Bill McCallum, one of the authors of the CCSSM, writes:

> "[F]luently" refers to how you do a calculation, whereas "know from memory" means being able to produce the answer when prompted without having to do a calculation. In the CCSSM, "fluent" means "fast and accurate."
>
> April 26, 2012

Developing both fluency and automaticity will be discussed further in this chapter.

What Are the Basic Multiplication Facts?

Basic facts are all the possible products of the digits 0 through 9 and the accompanying quotients. Often the products of 10 are included in this list because they play a key role in understanding multiplication of double-digit numbers and can be useful for deriving the "nine facts." Students should have quick recall of these basic multiplication facts. Because division facts can be easily derived when students understand the inverse relationship between multiplication and division, instruction should focus on developing automaticity with multiplication facts.

 Go To For more about the inverse relationship between multiplication and division go to Chapter 4: The Role of Concepts and Properties and Chapter 7: Developing Whole Number Division.

CCSSM and Math Facts

The CCSSM expectation states that by the end of grade 3 students will know from memory all products of two one-digit numbers. That is, the CCSSM expects students to focus on understanding the meaning for, and finding products of, single-digit multiplication and related quotients. To know from memory is not the same as memorization. Students can come to "know" their facts from memory in various ways, not necessarily through memorization. Although achieving fluency with single-digit facts is time consuming because there are many strategies and relationships that must be developed towards this end, it is manageable.

A Manageable Task

One aspect that makes learning multiplication facts so daunting is that students, teachers, and parents often think there are an overwhelming number of facts. Their perception is these facts are isolated bits of information to be recalled on demand and there is a seemingly endless list of them. When the approach to learning facts is based on memorization, this is the case; but in truth the task is much more manageable. If one considers the math facts from 0–10 then there are 121 multiplication math facts to learn. However, by applying the identity property of multiplication, the commutative property, and multiplication by 0, the number is quickly reduced to fewer than 50 unique math facts.

The commutative property shows us that once we have learned one fact we also know the fact partner to that fact (i.e., 3×4 is the partner to 4×3). Students will often discuss the relationship between 3×4 and 4×3 as the opposite, or the flip, of each other. Another common term used in classrooms is *turn-around facts*. Teachers can help link these student-invented phrases with the more precise language of commutative property in order to expose students to accurate math terminology.

Understanding the commutative property allows for the reduction of the total number of math facts students need to learn by almost half. For example, 3×4 and 4×3 can be considered one multiplication fact. Table 9.1 illustrates this idea. Each math fact shaded in gray has a partner fact in the nonshaded portion of the chart. This decreases the number of separate combinations from 121 to 66.

The multiplication facts where one factor is zero provide another opportunity to reduce the total number of math facts that must be learned. Multiplication by 0 should be discussed as $0 \times a = 0$, which means 0 groups of a. In using this definition of multiplication students begin to picture no groups of a number and therefore no objects to be accounted for. This is a straightforward set of facts for students to recall once students understand the concept behind multiplication by zero. Removing facts that have 0 as a factor (the first row of Table 9.1) further reduces the number of math facts to be learned to 55.

Table 9.1 Multiplication math fact table. The shaded section indicates the facts understood through the commutative property.

X	0	1	2	3	4	5	6	7	8	9	10
0	0	0	0	0	0	0	0	0	0	0	0
1	0	1	2	3	4	5	6	7	8	9	10
2	0	2	4	6	8	10	12	14	16	18	20
3	0	3	6	9	12	15	18	21	24	27	30
4	0	4	8	12	16	20	24	28	32	36	40
5	0	5	10	15	20	25	30	35	40	45	50
6	0	6	12	18	24	30	36	42	48	54	60
7	0	7	14	21	28	35	42	49	56	63	70
8	0	8	16	24	32	40	48	56	64	72	80
9	0	9	18	27	36	45	54	63	72	81	90
10	0	10	20	30	40	50	60	70	80	90	100

The identity property of multiplication helps to illustrate multiplication by one; $1 \times a = a$, or $a \times 1 = a$. Simply stated, the multiplicative identity property says that any time you multiply a number by 1, the product is that original number. Because students easily learn these facts, we can decrease the overall count by 10 (the second row of Table 9.1), leaving a total of 45 facts to develop with fluency and automaticity.

 Go To For an in-depth discussion of the commutative and identity properties of multiplication go to Chapter 4: The Role of Concepts and Properties.

Helping students see that there is a finite list of multiplication math facts to learn can decrease anxiety and frustration related to recall of math facts. Because 45 basic math facts remain to be learned with automaticity, we must consider how to accomplish this goal with a focus on sense making, reasoning, and relationships.

What Does It Mean to Know Your Basic Math Facts?

In the past, mastery of math facts was taught as drill and practice. Students were expected to memorize the basic multiplication facts, with teachers using a series of timed tests to both monitor students' progress and provide practice. This strategy of employing timed tests has been linked to math anxiety and to students turning away from mathematics (Boaler, 2015). Instruction focused on memorization and the use of drill alone is not the best way to help students become flexible thinkers of mathematics (Cook & Dossey, 1982; NRC, 2001). However, effective practice as a follow-up to instruction that highlights reasoning and strategies can strengthen memory and retrieval for many children.

Today it is generally agreed that learning math facts is a developmental process with instruction initially focused on building the concept of multiplication, then moving to fluency, or deriving facts through reasoning strategies, and ultimately to automaticity, or quick recall of the facts. There is evidence that learning math facts using strategies leads to higher accuracy and better transfer to unfamiliar math facts (Delazer et al., 2005; Kamii, 1994). It also builds strategies that can be relied upon as students work with larger numbers, as well as estimation for multiplication and division.

Building a Foundation on Conceptual Understanding

Throughout this book we have written about the important skills and concepts that a student must understand in order to be a strong multiplicative reasoner. We have emphasized the importance of focusing initial instruction on building an understanding of the many-to-one relationship, unitizing, and using visual models to represent multiplication situations. This is equally true for the development of fluency and automaticity of math facts. Understanding is the first step toward math fact acquisition, but it is not sufficient for assuring automaticity with the multiplication facts. Indeed, researchers indicate that learning basic math facts takes time, and focusing on memorization prior to students having enough experiences with developing strategies for deriving basic facts can have a negative impact on both the students' conceptual understanding and number sense (Van de Walle et al., 2012).

One should not expect students to memorize math facts until they have had multiple experiences with multiplication and division, both in and out of context and with the use of concrete models. The next section is focused on describing and illustrating strategies for deriving math facts that are focused on developing fluency.

Using Properties of Operations and Reasoning Strategies to Build Fluency

Properties of operations and visual models can be used to build conceptual understanding and fluency with multiplication facts. Instruction should be focused on developing understanding by creating a network of relationships that build on each other and can then be leveraged towards automaticity (e.g., Baroody, 1999).

 The *OGAP Multiplication Progression* can be used to help develop fluency and automaticity with math facts just as it is used to develop multidigit multiplication and division understanding and fluency. In particular, fluency can be built by drawing on the strategies at the

Transitional and *Multiplicative* levels. Reasoning strategies provide students with tools to move from knowing very few multiplication facts by memory to deriving facts with relative ease.

Notice that in the *Additive* level, strategies are reliant on adding or counting all. Although this is often where initial instruction related to multiplication will begin, these strategies will not provide students with fluency for deriving multiplication facts. Relying too long on additive strategies can result in challenges of efficiency and accuracy for many students as they encounter more complex problems, as evidenced in Figure 9.1. The evidence in Milo's work shows that instead of easy access to the math fact 6 × 7 when using the traditional US algorithm he used repeated addition and building up.

Figure 9.1 Milo's response. Milo uses the traditional US algorithm to find the answer to 273 × 6 but has to use addition to find 6 × 7.

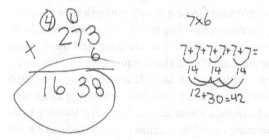

At the *Transitional* level you will notice that students use a variety of strategies such as skip counting. The use of skip counting begins in many kindergarten classrooms, particularly for the factors 2, 5, and 10. Students become familiar with the rhythm and order of these series of numbers. This knowledge can be built upon to strengthen understanding of math facts for multiplication. The link between skip counting and multiplication depends on whether or not students have mental images of the accumulating composite units (2 cookies to a package) with the correct quantity and the correct number name, as shown in Figure 9.2. This figure illustrates the connections between the mental image of the number of objects (cookies), the skip counting pattern with the accumulating composite units, and the multiplication math fact 4 × 2 = 8

Figure 9.2 An illustration of the skip counting pattern representing 4 × 2 = 8.

The use of skip counting based on an understanding of accumulating units can help students recall the basic multiplication facts for 2's, 5's and 10's with relative ease.

Another strategy evident in the *Transitional* level is the use of area and open area models. A student's ability to sketch area and open area models to make sense of multiplication and derive math facts is invaluable. Remember, in the context of area, all math facts create a rectangle. That is, the two factors are represented by the dimensions of the two sides of a rectangle, and the product is represented by the area of the same rectangle. Thus 3 × 6 is represented by a rectangle with side lengths or dimensions of 3 and 6, as shown in Figure 9.3. As students are developing strategies to make sense of basic multiplication facts, area models are particularly useful in helping them create mental images of the factors and products.

Figure 9.3 Ryan's response. Ryan sketched an area model for finding the product of 3 × 6.

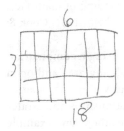

Knowledge and use of the distributive, associative, and commutative properties is at the *Multiplicative* level of the progression. Having flexible use of area and open area models and an understanding of properties of operations provides students with the skills and knowledge to derive unknown math facts and move toward fluency and automaticity. Figure 9.4 is an example of how students can use known facts, open area models, and understanding of properties of operations to derive unknown facts. Both examples in Figure 9.4 illustrate the use of the "known five facts" along with a flexible understanding of the distributive property. We can think of a known fact as an anchor fact. Anchor facts can be used to derive other facts. In the example on the left in Figure 9.4 the anchor fact of 6 × 5 is used to derive the product of 6 × 8.

Figure 9.4 Examples of possible open area models for deriving the math fact 6 × 8 = 48.

5 + 3		8	
6	30	18	

$(6 \times 5) + (6 \times 3) = 48$

5	40
+1	8

$(5 \times 8) + (1 \times 8) = 48$

Notice that the drawings of 6 x 8 represent the relative proportionality of the dimensions, resulting in partial areas that also represent their relative magnitude, helping students make sense of the math facts they are working with. Keeping one dimension stable and distributing across the other dimension when using open area models for developing fluency with math facts makes this a flexible strategy.

Ultimately the goal is for students to understand the properties and relationships well enough to derive facts without needing to sketch an area model to see what is happening. So a student who does not know the product of 6 and 8 could think, $5 \times 8 = 40$ and I still need to find 1×8 to get the answer to 6×8. This kind of flexible strategy relies on students having multiple experiences drawing area models and explicit instruction linking understanding of models and properties to deriving unknown math facts. Research indicates that student achievement of single-digit math facts improves with practice only after students have had many opportunities to build understanding of math facts through experiences with fluency strategies (Carnine & Stein, 1981; Cook & Dossey, 1982; Rathmell, 1978).

All of the 45 remaining math facts can be derived using a combination of the strategies, including using known facts, visualizing area and open area models, and an understanding of the distributive and associative properties of multiplication. A number of other relationships and strategies can make some facts easier than others to remember. One example is doubling.

Students' knowledge of doubles facts in addition can be used to build the idea of multiplication because knowing that multiplying by 2 is the same as adding two of the same number allows students to tie new knowledge, multiplying by 2, to previous knowledge, adding the same number (e.g., $6 + 6 = 2 \times 6$). This idea is important and *allows students to extend their understanding to unknown situations of multiplying by 2* (Baroody, 1999). Doubling, or multiplying by 2, can also be applied to facts that involve multiplication by 4. For example, 4×8 can be thought of as (2×8) doubled, or $(2 \times 8) + (2 \times 8)$. Generally, doubles are easy for children to learn and remember.

Table 9.2 summarizes the strategies and properties that were discussed in this chapter to help students build fluency that can lead to automaticity. As students have meaningful experiences using these strategies, they become more fluent and approach automaticity with their math facts. In addition, using properties of operations in this way provides multiple interactions and a deeper understanding of those properties, which play such a key role in multiplicative reasoning for years to come.

Table 9.2 Math fact strategies and properties discussed in this chapter.

Strategy	Used to Derive	Examples
Commutative property	Turn-around facts	$6 \times 7 = 7 \times 6$
The identity property of multiplication	All facts with 1 as a factor	$6 \times 1 = 6$ $1 \times 6 = 6$
Doubling	Facts with an even number as one of the factors	$6 \times 2 = 6 + 6$ $6 \times 4 = (6 \times 2) \times 2$
Skip counting	Facts in which a skip counting pattern of one of the factors is known	$6 \times 5 = 30$ 5, 10, 15, 20, 25, 30
Area model	Any fact by making a sketch using an understanding of the relationship between the dimensions and area of a rectangle with the ability to see and count all the squares	*(sketch of a rectangle, 6 wide and 3 tall, with grid squares, labeled 18)*
Open area model	Any fact by making a sketch based on an understanding of the relationship between dimensions and area of a rectangle and the distributive property.	*(sketch of a rectangle split into 5 + 3, height 6, with 30 and 18 inside)* $(6 \times 5) + (6 \times 3) = 48$
Distributive property and anchor facts	Any fact by using a known or anchor fact and the distributive property	$6 \times 9 = (5 \times 9) + (1 \times 9)$ 5×9 is the anchor fact
Inverse relationship between multiplication and division	Division facts	$54 \div 6 = ?$ $6 \times ? = 54$

From Fluency to Automaticity

As indicated at the beginning of the chapter, quick recall of basic math facts for multiplication is critical for a student's future success in mathematics. Once students understand and can use fluency strategies, as discussed earlier, it becomes important for them to practice math facts so they become automatic.

We will discuss two different kinds of fact practice, which we will refer to as targeted and general practice. Targeted practice is an opportunity to practice those facts identified as not automatic or fluent for a specific student. General practice is for ongoing exposure and practice with all multiplication facts. Both kinds of practice are important and are discussed in detail next. Practice should occur daily while students are trying to develop fluency and automaticity with math facts. Alternating between targeted and general fact practice on a daily basis provides a variety of practice and promotes student engagement. The time needed for mastery of facts can vary greatly from student to student.

Practicing for Automaticity

It is important to remember that practice or drill for automaticity is only effective as a follow-up to the fluency work done through a focus on strategies and reasoning. Because the work done on fluency strategies builds on relationships between the math facts, it goes without saying that practicing for quick recall should give opportunity for students to practice a mixed set of facts.

Targeted Fact Practice

Targeted practice can be thought of as strategic practice because it is focused on building fluency with the specific facts a student does not know. The first task of setting up targeted fact practice is to identify the math facts students know and do not know. Strategies for gathering this information about the students in your classroom might include using responses from group discussions, paper-and-pencil activities, and individual interviews. This information is essential for the teacher because future instruction should be guided by students' prior knowledge of specific math facts. Discussed next find two targeted strategies: math fact interviews and flash cards.

Math Fact Interviews: One way to gather information is to interview a student using a set of flashcards with no replications. For example, the teacher would remove the 4×6 card and leave the 6×4 card in the set. Have a student sit across from you and show them each card, sorting them into three piles as they give an answer: 1) knows automatically (within 3 seconds), 2) fluent (derives within 3 to 5 seconds), and 3) does not know. For those facts students know automatically or within 3 seconds without calculating, put them in the "knows automatically" pile. For those facts they have a strategy that allows them to derive the fact with relative ease and efficiency (within 3 to 5 seconds), place them in the "fluent" pile, and for those facts that do not go in either of these piles put them in a third pile we will call "does not know." The cards in the "does not know" pile will be those facts most difficult for the student to learn because they do not have an efficient strategy to derive the fact.

Once all the cards are in piles use a chart like the one in Figure 9.5 to indicate the student's knowledge of the multiplication facts. In Figure 9.5 the teacher crossed out the facts Sofia knew automatically, circled the facts she had a fluent strategy for, and left the facts blank Sofia did not know. As you can see, Sofia has a fluent strategy for 12 facts. The goal will be for her to practice those facts to move them from her fluent pile to her automatic pile. The teacher will help Sofia develop strategies for each fact she does not presently know so they can be moved into the fluent pile for practice.

Figure 9.5 Math fact information for Sofia.

~~1×1~~	~~1×2~~	~~1×3~~	~~1×4~~	~~1×5~~	~~1×6~~	~~1×7~~	~~1×8~~	~~1×9~~	~~1×10~~
	~~2×2~~	~~2×3~~	~~2×4~~	2×5	~~2×6~~	(2×7)	~~2×8~~	(2×9)	~~2×10~~
		~~3×3~~	(3×4)	3×5	(3×6)	(3×7)	~~3×8~~	3×9	~~3×10~~
			(4×4)	~~4×5~~	(4×6)	4×7	(4×8)	(4×9)	~~4×10~~
				~~5×5~~	(5×6)	~~5×7~~	~~5×8~~	~~5×9~~	~~5×10~~
					6×6	6×7	6×8	(6×9)	~~6×10~~
						7×7	7×8	~~7×9~~	~~7×10~~
							8×8	8×9	~~8×10~~
								~~9×9~~	~~9×10~~
									~~10×10~~

Flash Cards: A familiar strategy is the use of flashcards to help with targeted fact practice. When students are trying to move from fluent to automatic, they can write the fact on one side and the product on the other side. For those facts that are in the "do not know" pile students can create visual model *clues* on the flashcards as one way to bridge from not knowing a fact at all to getting a clue and not just turning the card over to get the answer. They can fall back on known facts and an understanding of area models and properties of operations to decide what the best strategy or clue is for them to derive unknown facts. Figure 9.6 is an illustration of what the front of a flash card might look like if the student is trying to develop a strategy for deriving the fact 6 × 6.

Figure 9.6 Sofia's sample flash card for 6 × 6. Sofia can construct a visual "clue" to help derive the math fact. On the back side of the flash card would be the product.

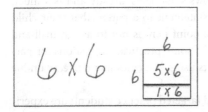

When students have free time in class they can practice with their set of personalized flash cards. At a designated period two to three times a week they should practice with a partner with the goal of moving towards automaticity. Although students may have a number of facts they need to practice, limiting the set of cards to no more than 10 will focus attention on a smaller, more manageable set of facts. The others can be attended to at a later time. The facts they practice should be a combination of those facts from the student's "fluent" pile and their "do not know" pile in order to vary the difficulty of the set.

General Fact Practice

General fact practice should be part of the regular experience for students as they acquire basic math facts. The purpose of general fact practice is to give students repeated exposure and practice with all math facts in a variety of situations. Math games are one way to encourage and vary practice of multiplication facts. A simple search for multiplication math games on the Internet will result in a long list of resources to choose from. A number of books have also been published containing a wide variety of games and activities for fact practice. Many math curriculum materials contain games for practicing math facts and incorporate the games into lessons for regular repeated practice. In addition, there are many apps for practicing math facts and free interactive games that can be accessed online. One website that has a number of good interactive games is NCTM Illuminations (*https://illuminations.nctm.org*). This site is managed and maintained by the National Council of Teachers of Mathematics.

Helping at Home

Practicing math facts is an important way we can involve parents and caregivers in their child's learning. But short of drilling math facts using either flash cards or worksheets, parents are often unsure how to support their child in learning their math facts. The more clarity we can provide about which facts their child needs to practice, the more likely it is they will see the job as doable. One way a teacher can accomplish this is to send home a chart like the one in Figure 9.5. Sharing with parents this level of detail about their child's math fact knowledge can be very beneficial as it provides a finite set of facts for them to focus on. It also acknowledges that their child knows some facts already and is a much clearer message than making a general statement to a parent that their child "needs to practice their math facts." The point here is not to assign drill and practice for homework, but rather to make some practical suggestions for parents to keep the focus on learning the facts using strategies that decrease undue stress and anxiety.

Parents can supplement the games and targeted practice students are experiencing at school with a kind of effortless fact practice at home. You can suggest that parents post facts their child needs to become automatic with around the house in places the student sees often. For example $8 \times 7 = 56$ can be written on a sheet of paper in large print and posted on the ceiling above their bed. Every time the child goes to bed they can look up and see both the fact and the answer. Posting the complete equation helps them to remember the math fact.

As you have seen, development of fluency and automaticity with math facts takes a multifaceted approach that incorporates development of fluency using strategies that develop understanding and then the implementation of targeted and general practice to achieve the goal of automaticity.

Chapter Summary

- Understanding is the first step in automaticity with math facts, but it doesn't necessarily lead to knowing the facts. Understanding should be followed up with explicit instruction focused on building strategies based on properties of operations and number relationships and then practice for automatic recall.
- Learning basic math facts takes time; focusing on memorization prior to having enough experiences with developing strategies for deriving basic facts can have a negative impact on both the students' conceptual understanding and number sense.
- It is important to remember that practice or drill for automaticity is only effective as a follow-up to the fluency work done through a focus on strategies and reasoning.
- Helping students derive unknown facts using known facts, area and open area models, and an understanding of properties of operations can enhance students' acquisition of math facts.
- There are two kinds of fact practice: targeted and general. Both are important in accomplishing the goal of automaticity with basic facts.

Looking Back

1. **The Commutative Property and Number Combinations in Multiplication:** A working understanding of the commutative property reduces the number of individual combinations a student has to learn.

    (a) Explain why this is so.

    (b) In what ways is 5×4 and 4×5 the "same" in:
       - An area model?
       - An equal groups model?

    (c) In what mathematical situations can one apply the commutative property? In what cases is the commutative property not applicable?

2. **Skip Counting and Number Combinations:**

    (a) Which number combinations might be first learned through skip counting?

    (b) How is skip counting different from knowing a multiplication "fact"? What does this suggest for instruction that supports building number combinations on skip counting?

    (c) How is skip counting by 4's related to skip counting by 2's? What other numbers besides 4 and 2 have this same relationship? How might one use these relationships to support students' learning number relationships?

3. **Strategies for Learning Specific Number Combinations:** Learning the 7's number combinations can be difficult for some students. Using ideas from this chapter identify strategies that one can use to assist learning the combinations for the 7's.

7 × 1	7 × 6
7 × 2	7 × 7
7 × 3	7 × 8
7 × 4	7 × 9
7 × 5	7 × 10

4. **The Relationship Between Multiplication and Division:** A solution to a division problem that is based on the relationship between multiplication and division is considered a *Multiplicative Strategy*. This relationship between multiplication and division can also help students learn the multiplication and division combinations together, rather than as separate and unconnected facts. List some ways students can connect the multiplication and division combinations for the expression 9 × 6.

Instructional Link

Use the following questions to analyze how your instruction and math program help students learn the basic multiplication number combinations. Consider the math program materials for previous grades if your grade-level materials do not specifically address basic multiplication number combinations.

- Is the instruction students receive related to number combinations primarily based on memorization or on strategies?
- Are there sufficient opportunities for students to develop strategies based on the commutative property, doubling, the relationship between multiplication and division, skip counting, and the relationship between multiplication and other operations?
- Does your program provide tools or suggestions for determining which number combinations students know and do not know?
- List ways you might include ideas from this chapter to address any shortcomings in your math program or in your instruction.

References

Baroody, A. J. (1999). The roles of estimation and the commutativity principle in the development of third graders' mental multiplication. *Journal of Experimental Child Psychology, 74*(3), 157–193.

Bass, H. (2003). Computational fluency, algorithms, and mathematical proficiency: One mathematician's perspective. *Teaching Children Mathematics, 9*(6), 322.

Battista, M. T. (2012). *Cognition-based assessment and teaching of multiplication and division: Building on students' reasoning (Cognition-based assessment and teaching).* Portsmouth, NH: Heinemann.

Battista, M. T., Clements, D. H., Arnoff, J., Battista, K., & Borrow, C.V.A. (1998). Students' spatial structuring of 2D arrays of squares. *Journal for Research in Mathematics Education, 29*(5), 503–532.

Beckmann, S. (2014). *Mathematics for elementary teachers with activities.* Boston: Pearson Education.

Bell, A., Fischbein, E., & Greer, B. (1984). Choice of operation in verbal arithmetic problems: The effects of number size, problem structure and context. *Educational Studies in Mathematics, 15*(2), 129–147.

Bell, A., Greer, B., Grimison, L., & Mangan, C. (1989). Children's performance on multiplicative word problems: Elements of a descriptive theory. *Journal for Research in Mathematics Education, 20*(5), 434–449.

Bell, A., Swan, M., & Taylor, G. (1981). Choice of operation in verbal problems with decimal numbers. *Educational Studies in Mathematics, 12*(4), 399–420.

Boaler, J. (2015). *Mathematical mindsets: Unleashing students' potential through creative math, inspiring messages and innovative teaching.* Hoboken, NJ: John Wiley & Sons.

Brown, M. (1981). Number operations. In K. Hart (Ed.), *Children's understanding of mathematics* (pp. 11–16). London: John Murray.

Brown, M. (1982). Rules without reasons? Some evidence relating to the teaching of routine skills to low attainers in mathematics. *International Journal of Mathematical Education in Science and Technology, 13*(4), 449–461.

Carnine, D. W., & Stein, M. (1981). Organizational strategies and practice procedures for teaching basic facts. *Journal for Research in Mathematics Education, 12*(1), 65–69.

Carpenter, T. P., Fennema, E., Peterson, P., Chiang, C. P., & Loef, M. (1989). Using knowledge of childrens thinking in classroom: An experimental study. *American Educational Reserach Journal, 26*(4), 499–531.

Carpenter, T. P., Franke, M. L., & Levi, L. (2003). *Thinking mathematically: Integrating arithmetic and algebra in elementary school.* Portsmouth, NH: Heinemann.

Carraher, T. N., Carraher, D. W., & Schliemann, A. D. (1987). Written and oral mathematics. *Journal for Research in Mathematics Education, 18*(2), 83–97.

CCSSO/NGA. (2010). *Common Core State Standards for Mathematics.* Washington, DC: Council of Chief State School Officers and the National Governors Association Center for Best Practices. Retrieved from http://corestandards.org.

Clarke, B. (2004). A shape is not defined by its shape: Developing young children's geometric understanding. *Journal of Australian Research in Early Childhood Education, 11*, 110–127.

Clements, D. H. (1999). Subitizing: What is it? Why teach it? *Teaching Children Mathematics, 5*(7), 400–405.

Clements, D. H., & Sarama, J. (2014). *Learning and teaching early math: The learning trajectories approach.* New York: Routledge.

Clements, D. H., Sarama, J., Spitler, M., Lange, A., & Wolfe, C. (2011). Mathematics learned by young children in an intervention based on learning trajectories: A large-scale cluster randomized trial. *Journal for Research in Mathematics Education, 42*(2), 127–166.

Cobb, P., Yackel, E., & Wood, T. (1988). Curriculum and teacher development: Psychological and anthropological perspectives. In E. Fennema, T. P. Carpenter, & S. J. Lamon (Eds.), *Integrating research on teaching and learning mathematics* (pp. 92–130). Madison, WI: Wisconsin Center for Education Research.

Cochran, K. F. (1991). Pedagogical content knowledge: A tentative model for teacher preparation. Paper presented at the annual meeting of American Educational Research Association, Chicago, IL.

Common Core Standards Writing Team. (2011). Progressions for the Common Core State Standards in Mathematics (draft). *K, counting and cardinality; K-5, operations and algebraic thinking.* Tucson, AZ: Institute for Mathematics and Education, University of Arizona.

Common Core Standards Writing Team. (2015). Progressions for the Common Core State Standards in Mathematics (draft). *Grades K-5, number and operations in base ten.* Tucson, AZ: Institute for Mathematics and Education, University of Arizona.

Cook, C. J., & Dossey, J. A. (1982). Basic fact thinking strategies for multiplication: Revisited. *Journal for Research in Mathematics Education, 13*(3), 163–171.

De Corte, E., Verschaffel, L., & Van Coillie, V. (1988). Influence of number size, problem structure, and response mode on children's solution of multiplication problems. *Journal of Mathematics Behaviour, 7*, 197–216.

Dehaene, S. (2011). *The number sense: How the mind creates mathematics.* Oxford: Oxford University Press.

Delazer, M., Ischebeck, A., Domahs, F., Zamarian, L., Koppelstaetter, F., Siedentopf, C. M.,. . . Felber, S. (2005). Learning by strategies and learning by drill: Evidence from an fMRI study. *Neuroimage, 25*(3), 838–849.

Ebby, C. B. (2005). The powers and pitfalls of algorithmic knowledge: A case study. *The Journal of Mathematical Behavior, 24*(1), 73–87.

Empson, S., & Levi, L. (2011). *Extending children's understanding of fractions and decimals.* Portsmouth, NH: Heinemann.

Fennema, E. H., Carpenter, T., Levi, L., Jacobs, V., & Empson, S. (1996). A longitudinal study of learning to use children's thinking in mathematics instruction. *Journal for Reserach in Mathematics Education, 27*(4), 403–434.

Fischbein, E., Deri, M., Nello, M. S., & Marino, M. S. (1985). The role of implicit models in solving verbal problems in multiplication and division. *Journal for Research in Mathematics Education, 16*(1), 3–17.

Forbringer, L., & Fahsl, A. J. (2009). Differentiating practice to help students master basic facts. In D. Y. White & J. S. Spitzer (Eds.), *Mathematics for every student responding to diversity: Grades prek-5* (pp. 65–74). Reston, VA: National Council of Teachers of Mathematics.

Fosnot, C., & Dolk, M. (2001). *Young mathematicians at work: Multiplication and division.* Portsmouth, NH: Heinemann.

Fuson, K. C., & Beckmann, S. (2012). Standard algorithms in the Common Core State Standards. *NCSM Journal, 14*(2), 14–30.

Gambrell, L. B., Koskinen, P. S., & Kapinus, B. A. (1991). Retelling and the reading comprehension of proficient and less-proficient readers. *The Journal of Educational Research, 84*(6), 356–362.

Graeber, A. O., & Tirosh, D. (1988). Multiplication and division involving decimals: Preservice elementary teachers' performance and beliefs. *The Journal of Mathematical Behavior, 7*(3), 263–280.

Greer, B. (1987). Nonconservation of multiplication and division involving decimals. *Journal for Research in Mathematics Education, 18*(1), 37–45.

Greer, B. (1988). Nonconservation of multiplication and division: Analysis of a symptom. *Journal of Mathematical Behavior, 7*(3), 281–298.

Greer, B. (1992). Multiplication and division as models. In D. Grouws (Ed.), *Handbook of research on mathematics teaching and learning* (pp. 276–296). Reston, VA: National Council of the Teachers of Mathematics.

Hart, K. M., Brown, M. L., Kuchemann, D. E., Kerslake, D., Ruddock, G., & McCartney, M. (1981). *Children's understanding of mathematics.* London: John Murray.

Hiebert, J., & Carpenter, T. P. (1992). Learning and teaching with understanding. In D. A. Grouws (Ed.), *Handbook of research on mathematics teaching and learning* (pp. 65–98). New York: Macmillan.

Hulbert, E., Petit, M., & Laird, R (2015). *OGAP Multiplicative Reasoning Professional Development.* Moretown, VT. Unpublished.

Kamii, C. (1994). *Young children continue to reinvent arithmetic: Implications of Piaget's theory.* New York: Teachers College Press.

Kamii, C. (1998). The harmful effects of algorithms in grades 1–4. In L. J. Morrow & M. J. Kenney (Eds.), *The teaching and learning of algorithms in school mathematics: 1998 yearbook* (pp. 130–140). Reston, VA: National Council of Teachers of Mathematics.

Kaput, J. (1989). Supporting concrete visual thinking in multiplicative reasoning: Difficulties and opportunities. *Focus on Learning Problem in Mathematics, 11*(2), 35–47.

Karp, K. S., Bush, S. B., & Dougherty, B. J. (2014). 13 rules that expire. *Teaching Children Mathematics, 21*(1), 18–25.

Kieren, T. E. (1988). Personal knowledge of rational numbers: Its intuitive and formal development. *Number concepts and operations in the middle grades, 2,* 162–181.

Kouba, V. L., & Franklin, K. (1993). Multiplication and division: Sense making and meaning. In R. J. Jensen (Ed.), *Research ideas for the classroom: Early childhood mathematics* (pp. 103–126). London: Macmillan.

Kouba, V. L., & Franklin, K. (1995). Multiplication and division: Sense making and meaning. *Teaching Children Mathematics, 1*(9), 574–578.

Lamon, S. J. (2005). *Teaching fractions and ratios for understanding: Essential content knowledge and instructional strategies for teachers.* New York: Routledge.

Lamon, S. J. (2012). *Teaching fractions and ratios for understanding: Essential content knowledge and instructional strategies for teachers.* New York: Routledge.

Lampert, M. (1986). Knowing, doing, and teaching multiplication. *Cognition and instruction, 3*(4), 305–342.

Lin, F.-L. (1991). Understanding in multiple ratio and non-linear ratio. *Proceedings of the National Science Council ROC(D), 1*(2), 14–30.

Markovits, Z., & Hershkowitz, R. (1997). Relative and absolute thinking in visual estimation processes. *Educational Studies in Mathematics, 32*(1), 29–47.

McCallum, B. (2012, April 26). *General Questions about the Mathematics standards.* Retrieved from http://commoncoretools.me/2012/04/02/general-questions-about-the-standards/

Morrow, L. M. (1985). Retelling stories: A strategy for improving young children's comprehension, concept of story structure, and oral language complexity. *The Elementary School Journal, 85*(5), 647–661.

National Research Council. (2001). *Adding it up: Helping children learn mathematics.* J. Kilpatrick, J. Swafford, & B. Findell (Eds.), Mathematics Learning Study Committee, Center for Education, Division of Behavioral and Social Sciences and Education. Washington, DC: National Academy Press.

Nesher, P. (1988). Multiplicative school word problems: Theoretical approaches and empirical findings. In J. Hiebert & M. Behr (Eds.), *Number concepts and operations in the middle grades* (pp. 19–40). Hillsdale, NJ: Erlbaum.

Nunes, T., & Bryant, P. (1996). *Children doing mathematics.* Malden, MA: Wiley-Blackwell.

Ongoing Assessment Project. (2005–2009). *OGAP questions and student work samples.* Unpublished manuscript.

Peterson, P. L., Carpenter, T. P., & Fennama, E. (1989). Teachers' knowledge in mathematics problem solving. *Journal of Educational Psychology, 8*(14), 558–569.

Petit, M. M., Laird, R. E., Marsden, E. L., & Ebby, C. B. (2015). *A focus on fractions: Bringing research to the classroom.* New York: Routledge.

Rathmell, E. C. (1978). Using thinking strategies to teach the basic facts. In M. Suydam & R. Reys (Eds.), *Developing computational skills, 1978 NCTM Yearbook* (pp. 13–38). Reston, VA: National Council of Teachers of Mathematics.

Reys, R. E., Lindquist, M., Lambdin, D. V., & Smith, N. L. (2009). *Helping children learn mathematics.* 9th edition. Hoboken, NJ: John Wiley & Sons.

Russell, S. J. (2000). Developing computational fluency with whole numbers. *Teaching Children Mathematics, 7*(3), 154.

Sarama, J., & Clements, D. H. (2009). *Early childhood mathematics education research: Learning trajectories for young children.* New York: Routledge.

Schwartz, J. L. (1988). Intensive quantity and referent transforming arithmetic operations. In J. Hiebert & M. J. Behr (Eds.), *Number concepts and operations in the middle grades* (pp. 41–52). Hillsdale, NJ: Erlbaum.

Shulman, L. S. (1986). Those who understand: Knowledge growth in teaching. *Educational Researcher, 15*(2), 4–14.

Siemon, D., Breed, M., & Virgona, J. (2005). From additive to multiplicative thinking: The big challenge of the middle years. Proceedings of the 42nd Conference of Mathematical Association of Victoria. Retrieved December 12, 2016, from www.eduweb.vic.gov.au/edulibrary/public/teachlearn/student/ppaddmulti.pdf.

Siemon, D., & Virgona, J. (2001). Roadmaps to numeracy: Reflections on the Middle Years Numeracy Research Project. Paper presented at Australian Association for Research in Education Conference, Fremantle, Perth.

Silver, E. A., Shapiro, L. J., & Deutsch, A. (1993). Sense making and the solution of division problems involving remainders: An examination of middle school students' solution processes and their interpretations of solutions. *Journal for Research in Mathematics Education, 24*(2), 117–135.

Sowder, J., Armstrong, B., Lamon, S., Simon, M., Sowder, L., & Thompson, A. (1998). Educating teachers to teach multiplicative structures in the middle grades. *Journal of Mathematics Teacher Education, 1*(2), 127–155.

Steffe, L. P. (1988). Children's construction of number sequences and multiplying schemes. In J. Hiebert & M. Behr (Eds.), *Number concepts in the middle grades* (pp. 119–146). Hillsdale, NJ: Erlbaum.

Steffe, L. P. (1992). Schemes of action and operations involving composite units. *Learning and Individual Differences, 4*(3), 259–309.

Steffe, L. P. (1994). Children's multiplying schemes. In G. Harel & J. Confrey (Eds.), *The development of multiplicative reasoning in the learning of mathematics* (pp. 3–39). Albany, NY: SUNY Press.

Tourniaire, F., & Pulos, S. (1985). Proportional reasoning: A review of the literature. *Educational Studies in Mathematics, 16*(2), 181–204.

Ulrich, C. (2015). Stages in coordinating units additively and multiplicatively. *For the Learning of Mathematics, 35*(3), 2–7.

Van de Walle, J., Karp, K., & Bay-Williams, J. (2012). *Elementary and middle school mathematics teaching developmentally*. New York: Pearson.

Vergnaud, G. (1983). Multiplicative structures. In R. Lesh & M. Landau (Eds.), *Acquisition of mathematics concepts and processes* (pp. 127–174). New York: Academic Press.

Vergnaud, G. (1988). Multiplicative structures. In J. Hiebert & M. Behr (Eds.), *Number concepts and operations in the middle grades, volume 2* (pp. 141–161). Reston, VA: National Council of Teachers of Mathematics.

Wong, M., & Evans, D. (2007). Improving basic multiplication fact recall for primary school students. *Mathematics Education Research Journal, 19*(1), 89–106.

Zweng, M. J. (1964). Division problems and the concept of rate: A study of the performance of second-grade children on four kinds of division problems. *The Arithmetic Teacher, 11*(8), 547–556.

About the Authors

Elizabeth T. Hulbert is an educational consultant in mathematics instruction and assessment and a co-director of OGAPMath LLC. Elizabeth provides leadership to the OGAP National Professional Development Team.

Marjorie M. Petit is an educational consultant in mathematics instruction and assessment and a co-director of OGAPMath LLC.

Caroline B. Ebby, Ph.D., is a senior researcher at the Consortium for Policy Research in Education (CPRE) and an adjunct associate professor at the Graduate School of Education at the University of Pennsylvania. Caroline is also a member of the OGAP National Professional Development Team.

Elizabeth P. Cunningham, Ph.D., is an assistant professor at the University of Michigan—Flint. Elizabeth is also a member of the OGAP National Professional Development Team.

Robert E. Laird is a research associate with the Vermont Mathematics Initiative (VMI) at the University of Vermont, co-director of OGAPMath LLC, and a member of the OGAP National Professional Development Team.

Index

absolute thinking 6

addition: additive strategies 19–21, 58, 139, 180; early additive strategies 19, 29, 57, 139; number relationships and action in 2; properties of operations in 64–70, *70–81*

additive strategies 19–21, 58, 139; math facts fluency and 180

algebraic expressions 162

algorithms 151–2; algebraic connections 162; Common Core State Standards for Mathematics and 153–4; division 162–8; linking open area models and 154–6, 160–1, 163; from models to standard 157–8; partial products 154–5, 157, 161; standard 157–8; traditional US 155–8, 161; without understanding 159–61

area and volume problem contexts 93–6, *108*

area models 45–6; open 46–7, 151, 154–6, 160–1, 163

arrays 43–5

associative property of multiplication *65*, 73–5

automaticity 176; moving from fluency to 183–4; practicing for 184–6

base-10 number system 59–60

Beckmann, S. 64, 157

binomials 162

Common Core State Standards for Mathematics 14, 176; algorithms 153–4; math facts 177; multiplication and division contexts 86–7; multiplicative patterns 99–100; OGAP Multiplication and Division Progressions and 30; procedural

fluency and 56; progression of numbers in problems 118

commutative property of multiplication *65*, 71–3, 177

concepts and properties: of operations 64–70, *70–1*; place value understanding 59; represented with visual models 71–9; role of place value in 60–4; unitizing 5, 20–1, 56–60

conceptual subitizing 48

conversion factors 108

dimensions as factors 45–6

distributive property of multiplication over addition *65*, 75–6, *77*

division 129–30; algorithms 162–8; contexts 86–7; interpreting remainders with 133–8; partitive and quotative 130–3; properties of operations in 64–70, *70–81*; word problem strategies 104

early additive strategies 19, 29, 139; unitizing and 57

early transitional strategies 21–2, 58, 139–40

engineering problems 122–3

equal groups *108*; problem contexts 87–9; using arrays to transition from 43–5; visual models 43

equal measures problem contexts 89, *108*

factors: as dimensions 45–6; magnitude of 112–15; three or more 117–18

fact practice 186

facts: basic multiplication 176; building foundation on conceptual understanding 179; Common Core